Emerging Contaminants in Water: Detection, Treatment, and Regulation

Emerging Contaminants in Water: Detection, Treatment, and Regulation

Editor

Frederick W. Pontius

MDPI • Basel • Beijing • Wuhan • Barcelona • Belgrade • Manchester • Tokyo • Cluj • Tianjin

Editor
Frederick W. Pontius
Civil Engineering and
Construction Management
California Baptist University
Riverside, CA
USA

Editorial Office
MDPI
St. Alban-Anlage 66
4052 Basel, Switzerland

This is a reprint of articles from the Special Issue published online in the open access journal *Water* (ISSN 2073-4441) (available at: https://www.mdpi.com/journal/water/special_issues/emerging_contaminants_detection_treatment_regulation).

For citation purposes, cite each article independently as indicated on the article page online and as indicated below:

LastName, A.A.; LastName, B.B.; LastName, C.C. Article Title. *Journal Name* **Year**, *Volume Number*, Page Range.

ISBN 978-3-0365-1857-2 (Hbk)
ISBN 978-3-0365-1858-9 (PDF)

Contents

About the Editor

Frederick W. Pontius is a professor at the Gordon and Jill Bourns College of Engineering, California Baptist University, Riverside, CA where he teaches courses in environmental engineering, fluid mechanics, water and wastewater treatment, project management, capstone design, and sustainable civil engineering. He has been a visiting professor at the Dept of Environmental Engineering, Chung Yuan Christian University, Taoyuan City, Taiwan. Prior to teaching he worked thirty years in industry as a consultant, environmental engineer, researcher, and as associate director regulatory affairs for the American Water Works Association. Dr. Pontius has published fourteen peer-reviewed papers, and over 200 technical articles. He has presented over fifteen multi-day training courses for USEPA and other agencies on water treatment and legislative and regulatory topics. Dr. Pontius has presented at the University of Colorado, Colorado School of Mines, the University of Michigan, the Harbin Institute of Technology, and the China Agricultural University.

Preface to "Emerging Contaminants in Water: Detection, Treatment, and Regulation"

Contaminants of emerging concern in water are an ongoing challenge globally. This Special Issue is devoted solely to emerging contaminants in water, bringing together recent research findings from leading scientists and engineers. Research papers in this special issue address sampling and analysis, treatment effectiveness, and regulation to limit human exposure. The hard work of the authors has provided an important contribution to the growing body of research on emerging contaminants in water.

Frederick W. Pontius
Editor

Editorial

Emerging Contaminants in Water: Detection, Treatment, and Regulation

Frederick Wendell Pontius

Department of Civil Engineering and Construction Management, Gordon and Jill Bourns College of Engineering, California Baptist University, 8432 Magnolia Ave., Riverside, CA 92503, USA; fpontius@calbaptist.edu

Contaminants of emerging concern in water are an ongoing challenge globally. This Special Issue is devoted solely to emerging contaminants in water, bringing together recent research findings from leading scientists and engineers. Emerging contaminants are naturally occurring, or synthetic chemicals or substances, recently detected or suspected to be present in water and whose toxicity or persistence pose some risk to human health or the environment. Contaminants of emerging concern are detected throughout the water cycle including surface waters, ground water, and effluents from wastewater treatment plants. The risks posed by their presence individually and as a mixture are not yet known. Significant attention is being given to developing analytical methods for their detection, evaluating treatment processes to remove them, and their regulation to limit human exposure.

Research papers in this special issue address a variety of emerging contaminants. Two studies address sampling and detection. Verlicchi and Ghirardini [1] investigate and compare the reliability of four different wastewater sampling techniques for pharmaceuticals and personal care products (PPCPs). Romero-Natale et al. [2] develop a spectrophotometric method for the determination of glyphosate based on complex formation between bis 5-phenyldipyrrinate of nickel (II) and the herbicide. The structure of this complex was elucidated.

Two studies address treatment topics. Murray and Örmeci's [3] report results of a bench treatment study on removal of microplastics and nanoplastics, Zbynek Hrkal et al. [4] conduct a two-year monitoring study of PPCPs at monthly intervals observing temporal changes in 81 substances in the source river and groundwater, evaluating bank infiltration and artificial recharge.

Two studies related to environmental waters are reported. Kubec et al. [5] evaluate possible effects of environmentally relevant concentrations (~1 $\mu g\ L^{-1}$) of two psychoactive compounds, venlafaxine and benzodiazepine oxazepam, on the behavior of freshwater crayfish. Putri et al. [6] applied multivariate statistical techniques and principal component analysis–multiple linear regression (PCA-MLR) to classify river pollution levels in Taiwan and identify possible pollution sources.

A review paper [7] completes this Special Issue examining the complex issues faced in developing regulations for perfluorooctanoic acid (PFOA) and perfluorooctane sulfonic acid (PFOS). PFOA and PFOS are receiving global attention due to their persistence in the environment through wastewater effluent discharges and past improper industrial waste disposal.

Citation: Pontius, F.W. Emerging Contaminants in Water: Detection, Treatment, and Regulation. *Water* **2021**, *13*, 1470. https://doi.org/10.3390/w13111470

Received: 23 April 2021
Accepted: 19 May 2021
Published: 24 May 2021

Funding: This research received no external funding.

Institutional Review Board Statement: Not applicable.

Informed Consent Statement: Not applicable.

Acknowledgments: The contributions and hard work of the authors of all papers in this Special Issue are greatly appreciated. Their work furthers our understanding of the contaminants studied and emerging contaminants in general. Thank you is also due to the MDPI Special Issue editors for their patience and persistence to make this special issue a success.

Conflicts of Interest: The author declares no conflict of interest.

References

1. Verlicchi, P.; Ghirardini, A. Occurrence of Micropollutants in Wastewater and Evaluation of Their Removal Efficiency in Treatment Trains: The Influence of the Adopted Sampling Mode. *Water* **2019**, *11*, 1152. [CrossRef]
2. Romero-Natale, A.; Palchetti, I.; Avelar, M.; González-Vergara, E.; Garate-Morales, J.L.; Torres, E. Spectrophotometric Detection of Glyphosate in Water by Complex Formation between Bis 5-Phenyldipyrrinate of Nickel(II) and Glyphosate. *Water* **2019**, *11*, 719. [CrossRef]
3. Murray, A.; Örmeci, B. Removal Effectiveness of Nanoplastics (<400 nm) with Separation Processes Used for Water and Wastewater Treatment. *Water* **2020**, *12*, 635. [CrossRef]
4. Hrkal, Z.; Eckhardt, P.; Hrabánková, A.; Novotná, E.; Rozman, D.; Hrabankova, A. PPCP Monitoring in Drinking Water Supply Systems: The Example of Karany Waterworks in Central Bohemia. *Water* **2018**, *10*, 1852. [CrossRef]
5. Kubec, J.; Hossain, S.; Grabicová, K.; Randák, T.; Kouba, A.; Grabic, R.; Roje, S.; Buřič, M. Oxazepam Alters the Behavior of Crayfish at Diluted Concentrations, Venlafaxine Does Not. *Water* **2019**, *11*, 196. [CrossRef]
6. Putri, M.S.A.; Lou, C.-H.; Syai'In, M.; Ou, S.-H.; Wang, Y.-C. Long-Term River Water Quality Trends and Pollution Source Apportionment in Taiwan. *Water* **2018**, *10*, 1394. [CrossRef]
7. Pontius, F. Regulation of Perfluorooctanoic Acid (PFOA) and Perfluorooctane Sulfonic Acid (PFOS) in Drinking Water: A Comprehensive Review. *Water* **2019**, *11*, 2003. [CrossRef]

Article

Removal Effectiveness of Nanoplastics (<400 nm) with Separation Processes Used for Water and Wastewater Treatment

Audrey Murray and Banu Örmeci *

Department of Civil and Environmental Engineering, Carleton University, Ottawa, ON K1S 5B6, Canada; audrey.murray@carleton.ca

* Correspondence: banu.ormeci@carleton.ca; Tel.: +1-(613)-520-2600 (ext. 4144)

Received: 12 January 2020; Accepted: 23 February 2020; Published: 26 February 2020

Abstract: Microplastics and nanoplastics are abundant in the environment, and the fate and impact of nanoplastics are of particular interest because of their small size. Wastewater treatment plants are a sink for nanoplastics, and large quantities of nanoplastics are discharged into surface waters through wastewater as well as stormwater effluents. There is a need to understand the fate and removal of nanoplastics during water, wastewater, and stormwater treatment, and this study investigated their removal on a bench-scale using synthesized nanoplastics (<400 nm) to allow controlled experiments. Plastic particles were created in the lab to control their size, and bench-scale dewatering devices were tested for their ability to remove these particles. Filtration with a 0.22 μm filter removed 92 ± 3% of the particles, centrifugation at 10,000 rpm (670,800 g) for 10 min removed 99 ± 1% of the particles, and ballasted flocculation removed 88 ± 3%. These results provide a general idea of the magnitude of the removal of nanoplastics with separation processes, and more work is recommended to determine the degree of removal with full-scale unit processes. Even though the removal was good using all three treatments, smaller particles escaping treatment may increase the nanoplastics concentration of receiving water bodies and impact aquatic ecosystems.

Keywords: nanoplastics; microplastics; wastewater; removal; settling; centrifugation; filtration

1. Introduction

Small plastic particles have become abundant in the environment due to the mass production of plastics and the low recycle rates for plastic products. Microplastics are defined as plastic particles having a size in the range of 100 nm to 5 mm. There is some disagreement amongst researchers as to the size of nanoplastics with some defining nanoplastics as those with at least one dimension less than 100 nm, and others defining nanoplastics as less than one micron [1–3]. The latter definition was adopted in this study. Larger microplastics that are in the mm size range are relatively easy to remove during water and wastewater treatment with settling and filtration-based processes, but plastics that are smaller than 1 μm may escape treatment in significant quantities in the effluent. Wastewater treatment plants are one of the main pathways of microplastic pollution, and treated wastewater discharges are an important source of nanoplastics, microbeads and synthetic textile fibers to surface waters [4,5]. Microplastic discharges from wastewater treatment plants have been reported to reach up to 15 million particles [4,5] and of particular concern for aquatic ecosystems are the nanoplastics in effluent discharges. Stormwater systems are also an important source of microplastics and provide a direct pathway to terrestrial microplastics into surface waters [6]. Stormwater ponds were reported to contain up to 22,894 microplastics/m^3 [7].

The prevalence, fate, and environmental effects of microplastics have been studied to some extent in recent years, but nanoplastics have been much less studied, due primarily to difficulties quantifying and characterizing nanoplastics in environmental samples. Microplastics are ingested by a wide range of fresh and saltwater species and can result in injury, disturbed feeding, disturbed swimming, immune responses, altered metabolism, tumors, and mortality among other effects [8]. There are very few studies on nanoplastics but they have been shown to have many of the same effects [3]. In fact, their small size, high surface curvature, and high surface area may increase their risk to wildlife. Their small size allows them to pass biological barriers and penetrate tissues, and they have been shown to interfere with algae photosynthesis [9].

Micro and nanoplastics can also pose a risk to aquatic environments through desorption of sorbed chemicals and/or transport of pathogens [8,10]. During wastewater treatment, nanoplastics can adsorb a wide range of chemicals resulting in particles with high concentrations of these chemicals which may later be desorbed after discharge into the environment. There is some debate over whether sorbed organic chemicals from microplastics are a significant threat to aquatic life, which are exposed to organic chemicals from other sources as well. However, bench-scale studies have shown adverse effects [11]. The larger surface area to volume ratio of nanoparticles increases their adsorption efficiency, but also increases the amount of adsorbed organic chemicals per particle if those particles escape treatment.

The fate of nanoplastics through wastewater treatment plant processes remains unknown. This is due to difficulties measuring nanoparticles in environmental samples. Methods for measuring nanoparticles, such as microscopy and spectroscopic tools, are time consuming and can disturb the characteristics of the particles [9,12]. Tools such as dynamic light scattering and nanoparticle tracking analysis can be used without disturbing the samples, but are unable to distinguish between plastic and naturally occurring nanoparticles [12]. The fate of microplastics in wastewater treatment has been more widely studied. For particles between 125 and 400 μm, Carr et al. [4] observed complete removal from wastewater treatment plants with tertiary gravity filtration, but incomplete (1 microplastic particle/1.14 thousand L) removal for plants with secondary treatment. Talvitie et al. [13] also found that although secondary treatment removed 98.4% of microplastics, the effluent had a fiber concentration 25 times and a particle concentration 3 times that of the receiving waters.

Due to difficulties in measuring nanoparticles in environmental samples, accompanied by the need to understand the fate of nanoplastics through treatment processes, this study investigated the removal of nanoplastics with gravity-based and mechanical separation processes on a bench-scale using synthesized nanoplastics (<400 nm). Physical separation methods were investigated and particle size profiles following treatment were used to provide additional information about the behavior of the particles. The selected separation processes are commonly employed for water, wastewater, and stormwater treatment, and the objective of the study was to determine and compare their effectiveness and further improve their performance in removing nanoplastics (<400 nm) from water.

2. Materials and Methods

This study investigated the removal of nanoplastic particles with common water and wastewater treatments, as well as the effects of these treatments on the particle size distribution of the remaining particles. Filtration, centrifugation, and ballasted flocculation were all evaluated. Uniform nanoplastic particles were prepared for this study, and turbidity was used to measure removal.

2.1. Overview

The nanoplastic particles used in this experiment were created in the lab using the procedure described in Section 2.2, below. They were then measured for size using scanning electron microscope (SEM) images as described in Section 2.3. Filtration (Section 2.6), centrifugation (Section 2.7), and ballast flocculation (Section 2.8) were all evaluated for their ability to remove the nanoplastic particles. Gravity settling tests were also conducted for comparison purposes (Section 2.9). Turbidity measurements (Section 2.4) were used to determine the removal efficiency for each of these processes. Zetasize measurements (Section 2.5) were used to determine changes in size distribution occurring during each removal process.

2.2. Preparation of the Nanoplastic Solution

Polymeric plastics were created containing carboxylic acid functional groups [14]. These plastics would be a good surrogate for polyethylene terephthalate (PET). Functional monomer methacrylic acid (MAA) (Sigma-Aldrich; Oakville, ON, Canada) and cross-linker ethylene glycol dimethacrylate (EGDMA) (Sigma-Aldrich; Oakville, ON, Canada) were dissolved in a porogen with a molar ratio of 1 mmol:8 mmol:6.7 mmol [15]. The porogen was composed of 40 mL of 1:3 (v:v) acetone (Fisher Scientific; Ottawa, Canada), and acetonitrile (Fisher; Ottawa, ON, Canada). 2% (w:w) of 2-isobutryonitrile (AIBN) was added as the initiator (Sigma-Aldrich; Oakville, ON, Canada). The mixture was mixed with a vortex mixer (Fisher Scientific Vortex Mixer, Chicago, IL, USA), deoxygenated with nitrogen for five minutes, and then placed in a 60 °C hot water bath for 24 h (Isotemp 220, Fisher, Chicago, IL, USA). The resulting nanoplastic particles were dewatered using a centrifuge (Thermo Scientific Sorval Legend RT+, Fisher Scientific, Chicago, IL, USA) at 10,000 rpm, air dried at room temperature, and ground manually.

Briefly, a nanoplastic solution was prepared by weighing 5 mg of nanoplastic particles and sonicating them (Vibracell Sonics, Sonics and Materials Inc., Newtown, CT, USA) in 1 L of deionized water for 15 min to disperse the plastic. The solution was stirred thoroughly before being dispensed.

2.3. Scanning Electron Microscope (SEM) Images

A Tescan Vegall XMU SEM instrument was used to obtain SEM images. The nanoplastics were coated with gold prior to imaging using RF (radio frequency) sputter (Anatech Hummer, Union City, CA, USA). The images were collected at a working distance of 7–8 mm.

2.4. Turbidity Measurements

The amount of nanoplastics removed was evaluated using before and after comparisons of turbidity. Turbidity measurements were taken using a Hach 2100 AN turbidity meter (Hach USA, Product Number: 4,700,100, Distributed by Hach Canada, London, ON, Canada).

Turbidity was chosen to measure particle removal because turbidity meters are most sensitive for nanoparticles with diameters close to the wavelength of visible light, and the nanoplastic particles had an average size of 217 ± 4 nm. A study with latex particles of varying sizes found that the maximum sensitivity occurred at a wavelength of 0.2 μm [16]. All turbidity measurements were higher than 0.5 NTU (nephelometric turbidity units) in this study. Additionally, a high initial particle concentration of 5 mg solids/L was chosen to enable the study of the behavior of nanoplastic particles through various water and wastewater treatments. At these concentrations, alternative methods such as counting under a microscope were not feasible; however, the turbidity was well above the 0.03 NTU detection limit identified by Gregory et al. [16], and turbidity presented a quick and straightforward means of measuring relative particle concentrations before and after treatment [16].

2.5. Particle Size Measurements

A Malvern Nano ZS Zetasizer (Malvern Instruments, St. Laurent, QC, Canada) was used to obtain particle size distributions and measure the average particle size before and after nanoplastics removal. Two experimental replicates, each consisting of two analytical replicates, were conducted.

2.6. Removal of Nanoplastics with Filtration

A bench-scale vacuum filtration set-up was used to evaluate the effectiveness of varying filter pore sizes on nanoplastic removal. The pore sizes investigated were: 0.22 μm (using a nylon syringe filter from Derian), 0.7 μm (GF/F glass microfiber filters, Whatman, Mississauga, ON, Canada,), 1 μm (GF/B glass microfiber filters, Whatmann, Mississauga, ON, Canada), 1.6 μm (G6 glass fiber circles, Fisher), and 3 μm (Grade 6 qualitative filters, Whatmann, Mississauga, ON, Canada). Three experimental replicates, evaluated with turbidity, were used for all removal experiments.

2.7. Removal of Nanoplastics with Centrifugation

Centrifuge time and speed were varied to determine optimum conditions for removal of nanoplastics. The nanoplastic solution was prepared as outlined above and poured into three 50 mL tubes. The centrifuge (Thermo Scientific Sorval Legend RT+, Fisher Scientific, Chicago, IL, USA) was run for varying times and speeds. Then a 20 mL sample was pipetted from the surface. In the first set of experiments, the centrifuge time was varied for a constant speed of 10,000 rpm corresponding to 670,800 times the force of gravity (g). In the second set of experiments, the centrifuge speed was varied for a constant time. Three experimental replicates, evaluated with turbidity, were used for all removal experiments.

2.8. Removal of Nanoplastics with Ballasted Flocculation

The jar test procedure for ballasted flocculation was followed as described by Desjardins et al. [17]. A jar test apparatus from Phipps and Bird (Richmond, WV, USA) was used with a flat blade impeller. The mixing speed was kept constant at 150 rpm throughout and the coagulants, microsand, and polymer were added as follows: (1) at time $t = 0$, the coagulant was added and the jar tester started; (2) at time $t = 2$ min, the microsand and a percentage of the polymer were added; (3) at time $t = 4$ min, the remainder of the polymer was added; (4) at time $t = 10$ min the mixing was turned off and the nanoplastics were allowed to settle by gravity; and (5) at time $t = 13$ min, a sample was taken approximately 5 cm below the surface. The times were varied in some cases to achieve optimum performance, but the order of the additions was maintained. The speed of impeller rotation for the jar-tester remained constant throughout. Three experimental replicates, evaluated with turbidity, were used for all removal experiments.

2.8.1. Chemicals and Sand for Ballasted Flocculation

The coagulant used for ballasted flocculation was aluminum sulfate (CAS 7784-31-8) (Anachemia, Montreal, QC, Canada). The polymer was Flopolymer CA4800 (SNF, Trois Rivieres, QC, Canada). Microsand samples were obtained from John Meunier Inc. (St. Laurent, QC, Canada).

2.8.2. Effect of Settling Time

To illustrate the effect that ballasted flocculation had on nanoplastic settleability, the nanoplastic solution was allowed to settle for 1 h following treatment, and the turbidity was monitored over time.

2.8.3. Control Parameters

Since there were many factors (alum dose, polymer dose, the percent of the polymer added initially, sand dose, mixing intensity, coagulation time, flocculation time, and settling time) capable of influencing the effectiveness of the ballasted flocculation treatment, a statistical approach employing a 2^{8-4} factorial design was used to determine which factors had a significant impact on performance. A linear regression model was obtained from the results.

2.8.4. Optimization of Ballasted Flocculation Conditions

Using the linear regression model obtained from the factorial design experiments, a path proportional to the coefficients in the regression model was followed until an increase in turbidity was observed. The conditions corresponding to the step before the increase in turbidity were taken as the optimum conditions.

2.9. Removal of Nanoplastics with Gravity Settling

To determine the settleability of the nanoplastic particles with no treatment, three 1 L beakers of the 5 mg/L nanoplastic suspension in tap water were monitored for 24 h. Samples were taken 5 cm below the surface of the water and measured for turbidity. Three experimental replicates, evaluated with turbidity, were used for all removal experiments.

3. Results and Discussion

3.1. Nanoplastic Characterization Prior to Treatment

Figure 1 shows a scanning electron microscope (SEM) image of the nanoplastic prepared for this study. From this image, the particles had an average diameter of 333 ± 76 nm. They were spherical in shape and allowed for well-controlled experiments.

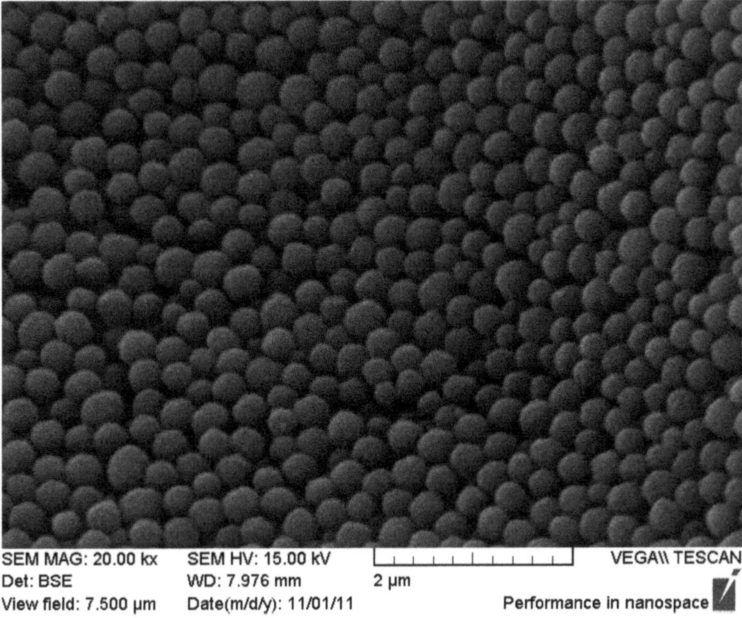

Figure 1. SEM (scanning electron microscope) image of nanoplastic particles.

To investigate the effect of various physical separation methods on removal and the size distribution of the nanoplastics, zeta size measurements were used to obtain a profile of the particle size distribution prior to treatment, and the results are shown in Figure 2. Zetasizer results show two analytical replicates. A second experimental replicate was also taken, and results were used to calculate average particle size, but are not shown.

The x-axis of Figure 2 shows the equivalent circle diameter, which is the diameter of the particles assuming the particles are spherical. Since the nanoplastics were roughly spherical, as shown in Figure 1, this was an accurate assumption. It is important to note that the x-axis is a logarithmic scale, so while the particle size distribution appears normal, it is skewed to the right. The y-axis shows the intensity of the signal for each diameter of particle, which provides a relative measure of how many particles of each diameter are present. The average particle size prior to treatment as measured by the Zetasizer was 217 ± 4 nm. This is smaller than the 333 ± 76 nm particle size obtained from the SEM image. The Zetasizer measures the hydrodynamic diameter, which is generally different from the geometric diameter. More importantly, the Zetasizer averages a much larger sample size than the SEM, where the particle size was estimated from the average diameter of 5 particles, and thus the Zetasizer may provide a better estimate.

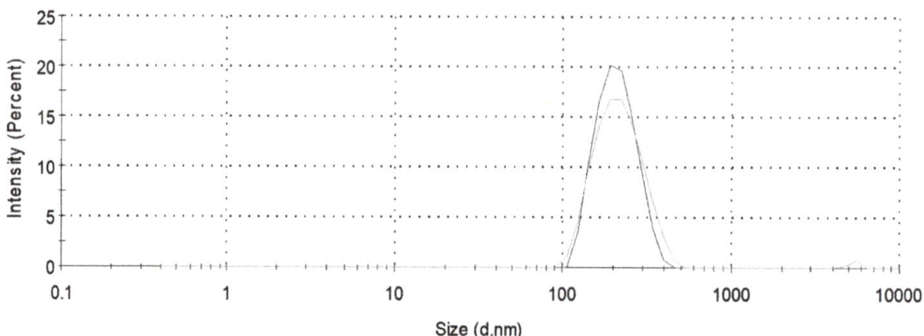

Figure 2. Particle size distribution of nanoplastic particles (Nano ZS Zetasizer). Lines indicate replicate results.

3.2. Filtration

Next, the effect of filtration on the concentration and size distribution of the nanoplastic particles was studied. Figure 3 shows the removal of nanoplastics using a bench-scale filtration set-up with filter pore sizes ranging from 0.22 to 3 μm. Turbidity measurements were used to determine the degree of removal.

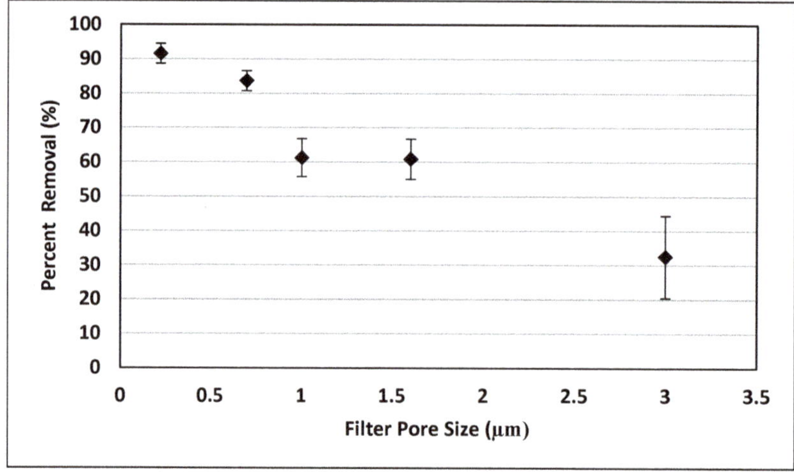

Figure 3. Percent removal of nanoplastics with filtration based on turbidity measurements.

As expected, the removal of the nanoplastics increased with decreasing pore size. The 3 μm filters removed 32 ± 12% of the particles, and the 0.22 μm filters removed 92 ± 3%. Since the nanoplastics had an average size of 333 ± 76 nm (from SEM images) or 217 ± 4 nm (from Zetasizer measurements), it was expected that the 0.22 μm filters would remove most but not all of the plastic particles. However, even the 0.7 μm filters were very effective at removing the particles and were capable of removing 84 ± 3%.

There are several reasons why filter sizes larger than the measured diameter of the particles might retain a fraction of the nanoplastics. Pore blocking from particles that have already been retained will reduce the effective pore size, or electrostatic charges may increase nanoplastic retention. The nanoplastics contained a negative surface charge under neutral conditions [14] and may have adsorbed to positive functional groups within the membranes [18]. Also, although the particles were fairly homogeneous in size and shape as seen from scanning electron microscope images, the particles may not have been completely separated in solution, and groups of particles could have been retained on much larger filter sizes.

The particle size distribution was measured following filtration with the 0.22 μm filter to gain a better idea of which particle sizes were removed during filtration. The results are shown in Figure 4. The average particle size measured by the zetasizer following filtration was 275 ± 70 nm. This is larger than the 217 ± 4 nm measured prior to filtration, although it was still within the range of error. This was not expected because filtration was expected to preferentially remove larger particles. It is possible that filtration caused agglomerations of finer particles that went through the filter leading to a larger average size. The difference between experimental replicates was also larger following treatment, which was expected due to slight differences in treatment experienced by samples during filtration. Comparing the distribution in Figure 4 to that in Figure 2, the spread of particle sizes was much smaller and there were fewer large particles. This also fits the hypothesis that filtration removed larger particles but caused the agglomeration of smaller particles by providing opportunities for particles to contact each other, reducing the spread in particle sizes. However, although the spread in particle sizes was reduced slightly, the distribution still appeared normal (although on a logarithmic axis this means it was skewed to the right) and the average particle size did not change significantly, indicating that the removal, which was significant (92 ± 3%) was fairly uniform across all particle sizes. This is noteworthy because preferential removal of larger particles during filtration could lead to an overly optimistic evaluation of filtration as a removal mechanism if smaller particles have a greater environmental impact. Full-scale removal would, of course, be different than bench-scale, but, due to difficulties measuring full-scale removal, lab experiments can provide a general idea of the behavior of nanoplastics during filtration. Additionally, 0.1 μm microfiltration or nanofiltration may be used at the full-scale to achieve greater removal.

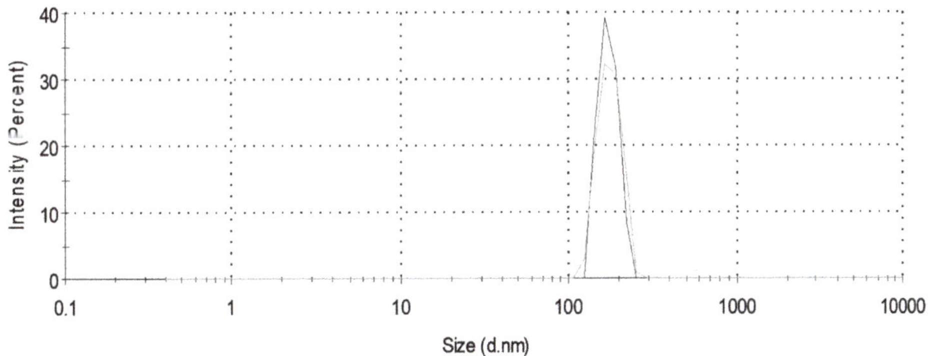

Figure 4. Particle size distribution of nanoplastic particles (Nano ZS Zetasizer) following filtration (pore size 0.22μm). Lines indicate replicate results.

3.3. Centrifugation

The effects of centrifugation on removal of nanoplastics was investigated for varying centrifuge times and speeds. The purpose was to give an idea of how many nanoplastics may be escaping centrifugation during wastewater treatment, in biosolids dewatering for example, where they would then re-join the wastewater influent. Figure 5a shows the percent removal for a constant centrifuge time (3 min) and varying centrifuge speed (5000–10,000 rpm), and Figure 5b shows the results for a constant centrifuge speed (10,000 rpm corresponding to 670,800 g) and varying centrifuge time (1–10 min). For a constant centrifuge time (Figure 5a) the percent removal increased distinctly for increasing centrifuge speeds between 5000 and 7000 rpm, from 49 ± 10% to 80 ± 7%. The increase was then more gradual for 8000–10,000 rpm from 84 ± 4% to 94 ± 6%. As presented in Figure 5b, the percent removal increased sharply for increasing centrifuge times from 0–3 min, increasing from 36 ± 7% after 1 min to 94 ± 6% after 3 min. The increase in percent removal was more gradual for longer centrifuge times, reaching 99 ± 1% for a 10-min centrifuge time.

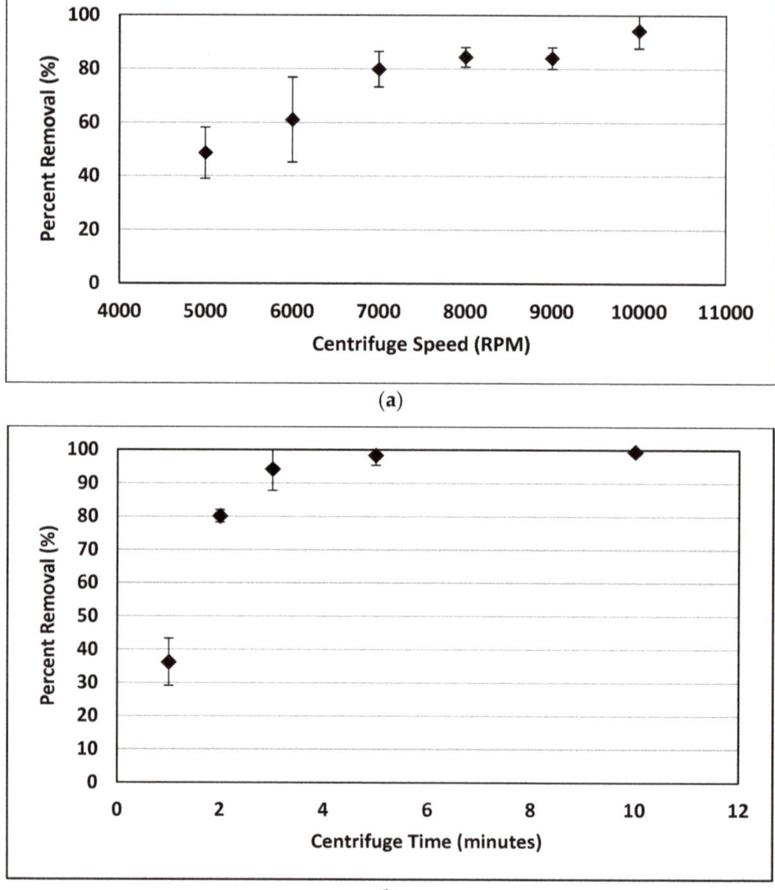

Figure 5. (**a**): Percent removal of nanoplastics by centrifugation at varying centrifuge speeds (time = 3 min) based on turbidity measurements. (**b**): Percent removal of nanoplastics by centrifugation at 10,000 RPM for varying centrifuge times based on turbidity measurements.

Although 99% removal appears excellent, any particles escaping treatment can increase the concentrations of nanoparticles in receiving waters. Murphy et al. [19] measured a decrease of 98.41% for microplastics in a secondary wastewater treatment plant, but noted that this still contributed 65 million microplastics per day to the receiving waters due to the large volume of wastewater treated per day.

Figure 6 shows the particle size distribution measured by the zetasizer following centrifugation at 10,000 rpm for 10 min. The average particle size following centrifuge treatment was 206 ± 45 nm. The error measures the difference between experimental replicates, which were themselves averages of two analytical replicates. Following centrifugation, there was a greater difference between experimental replicates than the untreated sample, presumably because there were more opportunities for differences to occur. In Figure 6, it appears as if the particle size distribution was skewed to the left. However, because the x-axis scale is logarithmic, the skew was, in fact, to the right. In comparison to the size distribution of the untreated sample shown in Figure 2, there are relatively fewer large particles (greater than 300 nm), and more smaller particles (less than 100 nm). The decrease in larger particles was expected because centrifugation can be expected to preferentially remove larger, heavier particles. An increase in small particles was not expected, but it is important to note that the Zetasizer measures intensity based on the relative particle count. Therefore, there may have been no actual increase in the number of smaller particles, but rather a decrease in larger and average size particles which led to an increase in the proportion of particles less than 100 nm. This may be of concern for wastewater treatment because it appears to indicate that smaller particles are not removed via centrifugation, and, as previously mentioned, smaller particles may have a greater environmental impact than larger particles.

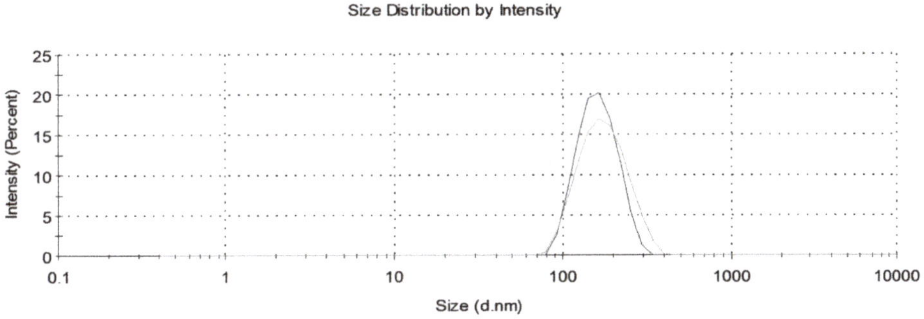

Figure 6. Particle size distribution of nanoplastic particles (Nano ZS Zetasizer) following centrifugation (time = 10 min). Lines indicate replicate results.

3.4. Ballasted Flocculation

Ballasted flocculation is typically used in drinking water treatment processes to remove fine particles; however, there are also applications in wastewater and combined sewer overflow treatment. Microsand, with a density of approximately 2700 kg/m^3, is incorporated into the coagulation-flocculation-settling regime typically used for water treatment to increase the size and density of particles to be removed [17]. Ballasted flocculation is effective in removal of small particles that are difficult to remove, and it was tested for removal of nanoplastics in this study.

Turbidity was used to measure the percent removal of nanoplastics, with no treatment over a 24 h period (Figure 7). Figure 8a then shows the results of extended settling following ballasted flocculation over the same period. The results shown represent an average of three replicates. For a 1-h period, with no treatment (Figure 7), there was no significant settling, and 0% removal is included in the range of error. Over a 24-h period, with no treatment the maximum removal with settling was 22 ± 8%. In contrast, Figure 8a shows that with ballasted flocculation a removal of 71 ± 5% was achieved after just 3 h, and removal had reached a steady state after 10 h. Figure 8b shows removal with ballasted flocculation for varying aluminum sulfate (alum) doses for a ten-minute settling period. When optimized for alum dose, ballasted flocculation could achieve 77 ± 15% removal after 10 min, and further extending the settling time (Figure 8a) did not significantly improve the performance. From these results, it is clear that ballasted flocculation can remove some nanoplastics from solution, but not effectively.

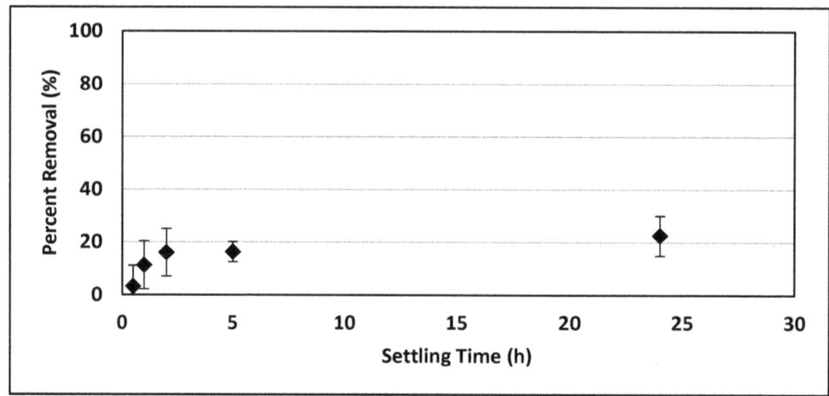

Figure 7. Percent removal of nanoplastics by gravity settling based on turbidity measurements (initial concentration = 5 mg/L).

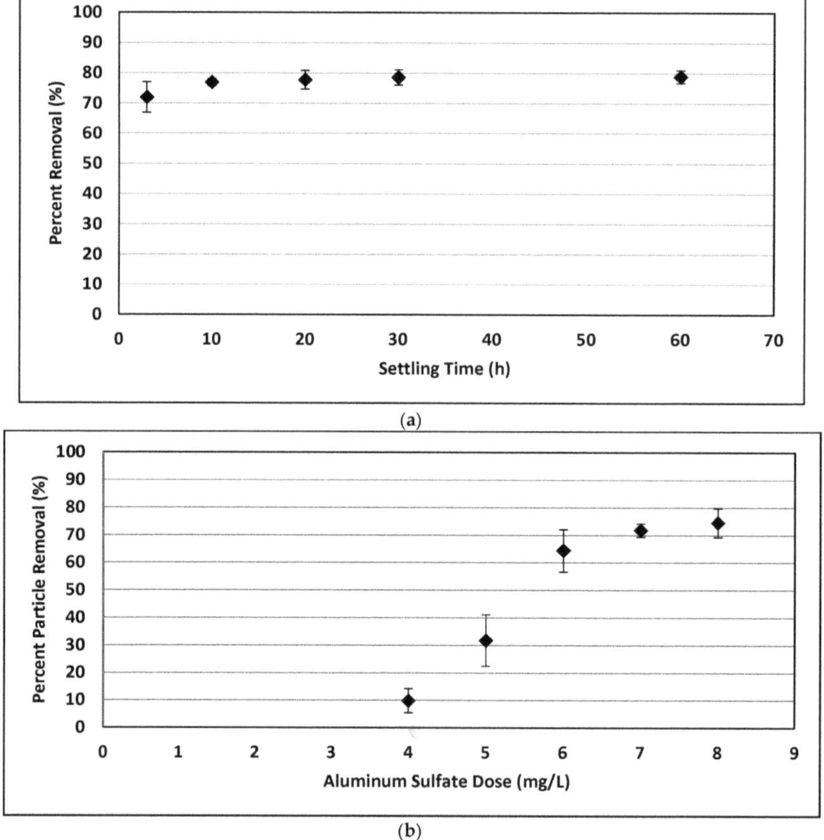

Figure 8. (**a**): Percent removal of nanoplastics by ballasted flocculation based on turbidity measurements (initial concentration = 5 mg/L; alum dose = 7 mg/L). (**b**): Percent removal of nanoplastics by ballasted flocculation with varying alum dose based on turbidity measurements (initial concentration = 5 mg/L; settling time = 3 min).

Unlike mechanical processes such as filtration and centrifugation, gravity-based settling processes are inexpensive, easy to employ, and practical in dealing with large flow rates. As a result, further experiments were conducted to optimize the ballasted flocculation conditions and improve the removal of nanoplastics. Since a ballasted flocculation process contains many design parameters which could be optimized a 2^{8-4} factorial design was used to determine the most significant parameters. This means that two levels were chosen for each of the eight parameters and varied in such a way that all the main effects could be determined without being confounded with two factor interactions, but higher order interactions were confounded. This design was chosen to minimize the number of tests required, while maximizing the results. All parameters related to process operation were considered, including alum dose, polymer dose, the fraction of the polymer added concurrently with the sand, sand dose, mixing intensity, coagulation time, flocculation time, and settling time. The alum dose, mixing intensity, and coagulation time were found to influence the final turbidity (refer to the Supplementary Materials for further details) of the water as described by the linear regression model shown in Equation (1). Equation (1) was then used to optimize nanoplastic removal as described in the Supplementary Materials. The optimum removal was found to be 88 ± 3%. The improved removal shows the importance of optimizing ballasted flocculation performance for removal of very small particles, but there was still a significant number of particles escaping treatment.

$$Turbidity = 2.52 - 0.35 alum\ dose - 0.049 mixing\ intensity - 0.26 coagulation\ time \quad (1)$$

It should be noted that optimum conditions are dependent on additional factors such as water characteristics or hydraulics, which are relevant for full-scale but not for jar tests and were not considered in this study. The optimum conditions determined with the jar-test procedure may not necessarily correspond to the optimum in a full-scale system but provide a general idea of the degree of removal possible. Nonetheless, Desjardins et al. [17] found good agreement between the performance of their jar-test procedure and three full-scale plants, with only a 7% difference.

Figure 9 shows the particle size distribution measured using the Zetasizer following ballasted flocculation. The average particle size was 251 ± 8 nm, which was larger than that for the untreated sample (217 ± 4 nm). This could be an indication of floc formation or it could be due to the presence of the ballast sand. Comparing the distribution shown in Figure 9 to that in Figure 2, it is difficult to see a difference visually. This indicates that there was probably only one type of particle present (nanoplastics), if there was a significant number of sand particles, the distribution would be expected to be bimodal.

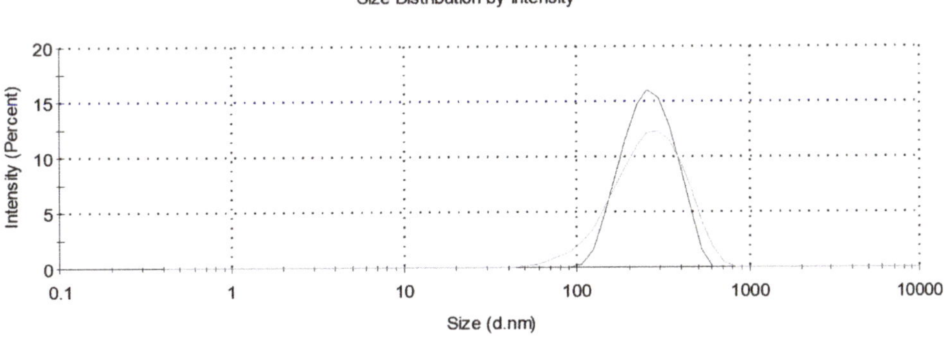

Figure 9. Particle size distribution of nanoplastic particles (Nano ZS Zetasizer) following ballasted flocculation. Lines indicate replicate results.

3.5. Comparison of Removal Processes

Figure 10 provides a bar-graph comparison of the best-case scenarios for ballasted flocculation, filtration, and centrifugation as options for the removal of nanoplastics from water samples following treatment as well as average scenarios for filtration and centrifugation, which were used to evaluate ballast flocculation as a pre-treatment. Since ballasted flocculation results in formation of larger flocs, a ballasted sand pre-treatment step was tested to investigate whether or not it could increase the efficiency of filtration. Since ballasted flocculation also increases the density of the flocs, it is possible that it could be used to decrease the centrifuge time or the centrifugal force required for nanoplastics removal and was also evaluated as a potential pre-treatment for centrifugation. The optimized ballasted flocculation conditions were used prior to filtration with a 0.7 µm pore size filter and centrifugation at 10,000 rpm for 3 min. The results showed there was no significant improvement in filtration or centrifugation performance after pre-treatment with ballasted flocculation.

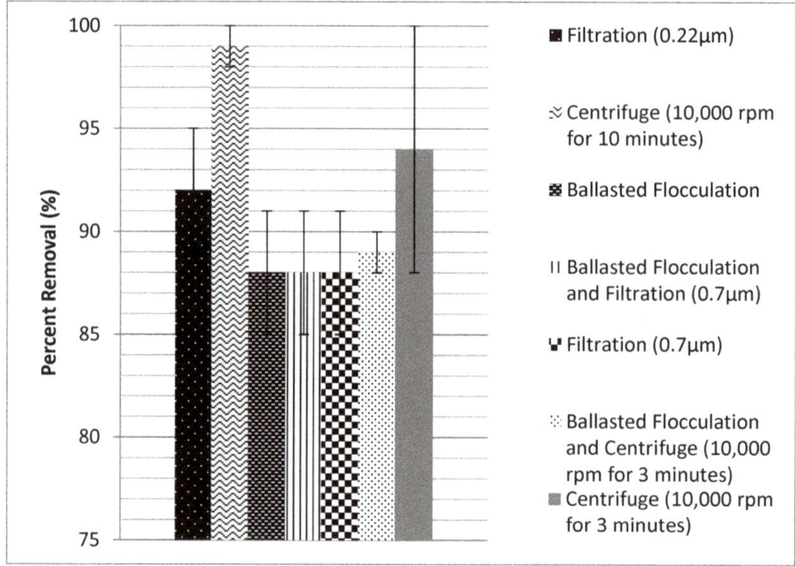

Figure 10. Comparison of percent removal of nanoplastics for various removal methods based on turbidity measurements.

An overall comparison of filtration, centrifugation, and ballasted flocculation either alone or as a pre-treatment, showed that centrifugation at 10,000 rpm for 10 min provides the best treatment and filtration with a 0.22 µm pore size filter was the second-best treatment option. The use of ballasted flocculation alone with gravity settling led to 88 ± 3% nanoplastics removal, but ballasted flocculation was not an effective pre-treatment prior to filtration or centrifugation.

4. Conclusions

This study investigated removal of nanoplastics (<400 nm) from water, at a bench-scale and using filtration, centrifugation, and ballasted flocculation. Filtration with 0.22 µm filters removed 92 ± 3% of the particles, and importantly did not show a size preference or change the distribution of the particles. Centrifugation at 10,000 rpm (670,800 g) for 10 min removed 99 ± 1% of the particles but did preferentially remove larger particles. This is a potential area of concern because smaller particles may have a greater environmental impact. Ballasted flocculation was able to remove 88 ± 3% of the particles. These results provide a general idea of the magnitude of removal of nanoplastics smaller than 400 nm with separation processes used for water, wastewater, and stormwater treatment. The removal was good using all three treatments, but it is important to note that even a small number of particles escaping treatment can increase the nanoplastics concentration of receiving water bodies. More work is recommended to determine the degree of removal with full-scale unit processes.

Supplementary Materials: The following are available online at http://www.mdpi.com/2073-4441/12/3/635/s1, Table S1-1: Factorial Experiment for Ballast Flocculation (data), Table S1-2: 2^{8-4} factorial design results and decoded design parameters for ballasted flocculation removal of nano-plastic particles, Table S1-3: ANOVA analysis of significant factors, Table S1-4: 2^{8-4} factorial design results and decoded design parameters for ballasted flocculation removal of nanoplastic particles, insignificant parameters removed, Table S1-5 ANOVA analysis of significant factors for alum dose, mixing intensity, and coagulation time, Table S2-1 Steepest ascent values used to optimized ballasted flocculation, Table S2-2: Steepest ascent optimization of ballasted sand flocculation.

Author Contributions: Conceptualization, B.Ö. and A.M.; methodology, A.M. and B.Ö.; investigation, A.M. and B.Ö.; validation, A.M.; formal analysis, A.M.; writing—original draft preparation, A.M.; writing—review and editing, B.Ö.; supervision, B.Ö.; funding acquisition, B.Ö. All authors have read and agreed to the published version of the manuscript.

Funding: This research was funded by the Natural Science and Engineering Research Council of Canada (NSERC) under the Discovery Program (RGPIN 06246).

Conflicts of Interest: The authors declare no conflict of interest.

References

1. Ng, E.-L.; Lwanga, E.H.; Eldridge, S.M.; Johnston, P.; Hu, H.-W.; Geissen, V.; Chen, D. An overview of microplastic and nanoplastic pollution in agroecosystems. *Sci. Total Environ.* **2018**, *627*, 1377–1388. [CrossRef] [PubMed]
2. Pico, Y.; Alfarhan, A.; Barcelo, D. Nano and microplastic analysis: Focus on their occurrence in freshwater ecosystems and remediation technologies. *Trends Anal. Chem.* **2019**, *113*, 409–425. [CrossRef]
3. Da Costa, J.P.; Santos, P.S.M.; Duarte, A.C.; Rocha-Santos, T. (Nano)plastics in the environment – sources, fates and effects. *Sci. Total Environ.* **2016**, *566*, 15–26. [CrossRef] [PubMed]
4. Carr, S.A.; Liu, J.; Tesoro, A.G. Transport and fate of microplastic particles in wastewater treatment plants. *Water Res.* **2016**, *91*, 174–182. [CrossRef] [PubMed]
5. Almroth, B.M.C.; Astrom, L.; Roslund, S.; Petersson, H.; Johansson, M.; Persson, N.-K. Quantifying shedding of synthetic fibers from textiles; a source of microplastics released into the environment. *Environ. Sci. Pollut. Res. Int.* **2018**, *25*, 1191–1199. [CrossRef] [PubMed]
6. Horton, A.A.; Walton, A.; Spurgeon, D.J.; Lahive, E.; Svendsen, C. Microplastics in freshwater and terrestrial environments: evaluating the current understanding to identify the knowledge gaps and future research priorities. *Sci. Total Environ.* **2017**, *586*, 127–141. [CrossRef] [PubMed]
7. Liu, F.; Olesen, K.B.; Borregaard, A.R.; Vollertsen, J. Microplastics in urban and highway stormwater retention ponds. *Sci. Total Environ.* **2019**, *671*, 992–1000. [CrossRef]
8. Eerkes-Medrano, D.; Thompson, R.C.; Aldridge, D.C. Microplastics in freshwater systems: A review of the emerging threats, identification of knowledge gaps and prioritisation of research needs. *Water Res.* **2015**, *75*, 63–82. [CrossRef] [PubMed]
9. Mattsson, K.; Hansson, L.A.; Cedervall, T. Nanoplastics in the aquatic environment. *Environmental Sciences: Processes and Impacts* **2015**, *17*, 1712–1721. Available online: https://pubs.rsc.org/en/content/articlelanding/2015/em/c5em00227c#!divAbstract (accessed on 25 February 2020).
10. Kirstein, I.V.; Kirmizi, S.; Wichels, A.; Garin-Fernandez, A.; Erler, R.; Martin, L.; Gerdts, G. Dangerous hitchhikers? Evidence for potentially pathogenic Vibrio spp. on microplastic particles. *Mar. Environ. Res.* **2016**, *120*, 1–8. [CrossRef] [PubMed]
11. Ziccardi, L.M.; Edgington, A.; Hentz, K.; Kulacki, K.J.; Kane Driscoll, S. Microplastics as vectors for bioaccumulation of hydrophobic organic chemicals in the marine environment: A state-of-the-science review. *Environ. Toxicol. Chem.* **2016**, *35*, 1667–1676. [CrossRef] [PubMed]
12. Paterson, G.; MacKenac, A.; Thomasa, K.V. The need for standardized methods and environmental monitoring programs for anthropogenic nanoparticles. *Anal. Methods* **2011**, *3*, 1461–1467. [CrossRef]

13. Talvitie, J.; Heinonen, M.; Paakkonen, J.-P.; Vahtera, E.; Mikola, A.; Setala, O.; Vahala, R. Do wastewater treatment plants act as a potential point source of microplastics? Preliminary study in the coastal Gulf of Finland, Baltic Sea. *Water Sci. Technol.* **2015**, *72*, 1495–1504. [CrossRef] [PubMed]
14. Lai, E.P.C.; De Maleki, Z.; Wu, S. Characterization of molecularly imprinted and nonimprinted polymer submicron particles specifically tailored for removal of trace 17b-estradiol in water treatment. *J. Appl. Polym. Sci.* **2010**, *116*, 1499–1508.
15. Wei, S.; Molinelli, A.; Mizaikoff, B. Molecularly imprinted micro and nanospheres for the selective recognition of 17beta-estradiol. *Biosens. Bioelectron.* **2006**, *21*, 1943–1951. [CrossRef] [PubMed]
16. Gregory, J. Turbidity and beyond. *Filtr. Sep.* **1998**, *35*, 63–67. [CrossRef]
17. Desjardins, C.; Koudjonou, B.; Desjardins, R. Laboratory study of ballasted flocculation. *Water Res.* **2002**, *36*, 744–754. [CrossRef]
18. Pontius, F.W.; Amy, G.L.; Hernandez, M.T. Fluorescent microspheres as virion surrogates in low-pressure membrane studies. *J. Membr. Sci.* **2009**, *335*, 43–50. [CrossRef]
19. Murphy, F.; Ewins, C.; Carbonnier, F.; Quinn, B. Wastewater Treatment Works (WwTW) as a Source of Microplastics in the Aquatic Environment. *Environ. Sci. Technol.* **2016**, *50*, 5800–5808. [CrossRef] [PubMed]

Article

Occurrence of Micropollutants in Wastewater and Evaluation of Their Removal Efficiency in Treatment Trains: The Influence of the Adopted Sampling Mode

Paola Verlicchi [1,2,*] and Andrea Ghirardini [1]

1 Department of Engineering, University of Ferrara, Via Saragat 1, 44122 Ferrara, Italy; andrea.ghirardini@unife.it
2 Terra & Acqua Tech Tecnopolo, University of Ferrara, Via Borsari 46, 44121 Ferrara, Italy
* Correspondence: paola.verlicchi@unife.it

Received: 13 May 2019; Accepted: 31 May 2019; Published: 31 May 2019

Abstract: The monitoring of micropollutants in water compartments, in particular pharmaceuticals and personal care products, has become an issue of increasing concern over the last decade. Their occurrence in surface and groundwater, raw wastewater and treated effluents, along with the removal efficiency achieved by different technologies, have been the subjects of many studies published recently. The concentrations of these contaminants may vary widely over a given time period (day, week, month, or year). In this context, this paper investigates the average concentration and removal efficiency obtained by adopting four different sampling modes: grab sampling, 24-h time proportional, flow proportional and volume proportional composite sampling. This analysis is carried out by considering three ideal micropollutants presenting different concentration curves versus time (day). It compares the percentage deviations between the ideal concentration (and removal efficiencies) and the differently measured concentrations (removal efficiencies) and provides hints as to the best sampling mode to adopt when planning a monitoring campaign depending on the substances under study. It concludes that the flow proportional composite sampling mode is, in general, the approach which leads to the most reliable measurement of concentrations and removal efficiencies even though, in specific cases, the other modes can also be correctly adopted.

Keywords: average daily concentration; mass loading; micropollutants; removal efficiency; sampling mode; uncertainties

---

1. Introduction

In planning a monitoring campaign, difficulties may arise in defining the sampling strategy, namely the mode and frequency of sample withdrawal in order to collect a number of samples which can be considered representative of the environment, the phenomenon or the process under study. Limiting attention to the water environment (namely raw wastewater, treated effluent, surface water and groundwater), different sampling modes may be utilized: water samples can be instantaneous (grab samples) or composite. In the second case, the resulting composite samples may be time proportional, flow proportional or volume proportional. Moreover, the reference interval for each composite sample could be 24 h or a fraction of the day (12 h, 4 h, or 3 h) [1]. With regard to withdrawal frequency, it is important to plan the sampling in order to pinpoint the (expected or potential) different behaviors in the occurrence of the compounds under study over a period of time [2,3].

In the case of monitoring campaigns tackling compounds occurring at very low concentrations, in the range of ng/L–µg/L—the so-called 'micropollutants'—it is fundamental to adopt an adequate sampling strategy and also to report it in detail along with the collected results [4–6]. Pharmaceuticals and personal care products, flame retardants and parabens are just some of the groups of (micro)pollutants of emerging concern. There has been a sudden increase in studies and publications dealing with the occurrence of these (micro)pollutants in different water environments, and relative removal technologies, from conventional treatments to the most promising technologies and different treatment trains. Most of them are still unregulated compounds (thus their limits in the case of discharge of a treated effluent into a surface water body have not yet been defined), but attention to their potential effects on the environment and human health is increasing and studies are in progress in many parts of the world [7–9].

Micropollutants can also be present in industrial wastewater. For instance, a petrochemical wastewater treatment plant may receive raw wastewater from different production wards within the industrial pole, characterized by a wide spectrum of pollutants. Cattaneo et al. [10] report the case of the petrochemical site of Porto Marghera, near Venice in Italy, where the purpose-built wastewater treatment plant must adhere to (strict) authorized limits for the occurrence of macropollutants (among them: suspended solids, biological oxygen demand, total Kjeldahl nitrogen, and nitrates) and ten micropollutants (the so-called "ten forbidden substances": cyanides, arsenic, cadmium, mercury, lead, organic chloride pesticides, hexachlorobenzene, tributyltin, polychlorinated biphenyls (PCB), dioxins and polycyclic aromatic hydrocarbons (PAH)) in the treated effluent. Sometimes, regulations may also require that the wastewater treatment plant guarantees removal for a selection of (micro)pollutants, in order to demonstrate that it acts as an efficient barrier against them. It is important to underline that in all these situations, a correct sampling mode must be adopted and clearly reported in detail with the results in order to be able to evaluate how representative and reliable the collected measured concentrations are.

Investigations into the occurrence of micropollutants in wastewater have highlighted that many of them may exhibit a substantial variation in concentration over the day (e.g., sulfamethoxazole and ciprofloxacin, [11–13]), the week (e.g., fluoruracil, diatrizoate, iomeprol and iohexol [2]), and the month (e.g., cefazolin and carbamazepine, [3]). Others have drawn attention to the temporal variation and distribution of selected pharmaceuticals in surface water bodies (among them [14,15]).

The issue of the influence of the sampling mode adopted in monitoring micropollutants has been addressed by many researchers in the last 10 years. Only in a few studies has this issue has been addressed with great detail (among them [2,4–6,16,17]); more often the issue is remarked on but not well discussed [1]. Particularly interesting are the sophisticated studies carried out by Ort and colleagues in [5,6,16,17] regarding the occurrence of pharmaceuticals and diagnostic agents in raw (municipal and hospital) wastewater and treated effluents, as well as in surface water, leading to suggestions for monitoring campaigns of micropollutants on the basis of the number of pulses containing the substance of interest (i.e., the number of toilet flushes at the sampling location) for a catchment area.

The current paper focuses on this issue following another approach: it faces the question by presenting and discussing numerical examples referring to some (representative) micropollutants characterized by different concentrations versus time curves.

In particular, it refers to three substances presenting very different profiles of concentration over the day (a highly variable compound, a randomly variable compound and a compound with low variability), and for each of them it evaluates: (i) the average daily concentration in the case of grab sampling, 24-h time proportional, 24-h volume proportional and 24-h flow proportional composite sampling; and (ii) the daily mass loading based on the estimated average concentrations and the provided flow rate. Finally, it assesses (iii) the removal efficiency for one of the three substances based on the different values of average concentrations found by applying the different sampling modes. This study ends with the evaluation of the (percentage) deviations between the "measured" concentration obtained by adopting a specific sampling mode and the "ideal" average concentration of each representative compound, as well as the (percentage) deviation between the evaluated removal efficiency and the ideal one.

2. Materials and Methods

This study refers to a "theoretical" case study regarding the occurrence of three micropollutants characterized by a different concentration profile versus time (over the day). The simulated substances do not correspond to three specific compounds, but each of them is representative of a group of compounds with a similar concentration trend versus time (see Section 2.1). In this context, the investigations by [11,12,18,19] clearly show the variations in the concentration of micropollutants in municipal raw wastewaters and hospital effluents over a typical day. These experimental values provide us useful insights into the different possible profiles of concentration of micropollutants and allow us to define theoretical ad hoc curves of concentrations versus time for three different representative scenarios.

As to flow rate, the study refers to a small urban settlement, which, according to the technical literature, is characterized by enhanced variations at well-known day hours [20]. A very similar flow pattern was found for the effluent of a medium-large hospital [12,21,22]. In this context, an ad hoc curve of flow rates versus time (during a typical day) was defined on the basis of literature data and evidences [20,21] (see Section 2.2).

It is important to keep in mind that, in the following, attention has to be paid to the variations in concentrations and flow rate over the day and not to the specific (absolute) values reported in the graphs. This means that considerations and results developed in this study can be applied to a small urban settlement as well as a medium to large hospital characterized by similar concentration profiles but different (maximum and minimum) concentration values (often higher in the hospital effluent, [3,21].

2.1. Definition of Representative Compounds

Three key compounds were considered for the study:

- a substance whose concentration in wastewater presents few but evident variations over the day, such as the diagnostic agents gadolinium and iopamidol [18], the cytostatic agent 5-fluoruracil [2] or the diuretic furosemide and the antibiotic sulphamethoxazole [13]. Such a substance is called a 'high variability substance', HV_Sub. During the night, its concentration decreases even lower than the corresponding limit of detection (Lod) for some hours;
- a substance whose concentration in wastewater presents a modest variation over the day, and is also detectable during the night, such as the anti-inflammatory ketoprofen [19], the antiseptic triclosan and the anticonvulsant agent phenytoin [13], and the antibiotic trimethoprim [13,23]. This is called the 'low variability substance', LV_Sub. It may happen that during the night its concentration decreases to values below its limit of detection, but only for very short periods;
- a substance whose concentration "randomly" varies over the day, such as the antibiotics ciprofloxacin [12] lincomycin [23], the anti-inflammatories diclofenac [13], and 4-tert octylfenol (a degradation product of a surfactant). This substance is called a 'random variability substance', RV_Sub. Its profile pattern is not easily predictable.

Based on literature data and in particular on the observed temporal variations in concentrations reported for the cited compounds in wastewater [2,4,11–13,18,19,23], 24 values of concentrations were set (one for each hour of a day) for the three key compounds (Table S1). Based on them, a nonlinear regression curve was carried out for each substance, by means of the software MATLAB R2018b. The corresponding polynomial functions are reported in Equations (1)–(3) (where concentration is in ng/L and time in min). In this way, the concentration c versus time t curves were set as continuous functions $c(t)$ (Figure 1).

$$c_{HV\_Sub}(t) = 0.015t^8 - 0.16t^7 + 6.58t^6 - 141.33t^5 + 1660.2t^4 - 10367t^3 + 30900t^2 - 32689t + 1289.9 \quad (1)$$

$$c_{RV\_Sub}(t) = -0.27t^5 + 14.93t^4 - 280.73t^3 - 1977.7t^2 - 2162.1t - 14009 \quad (2)$$

$$c_{LV\_Sub}(t) = +0.008t^9 + 0.2t^8 - 3.07t^7 + 26.93t^6 - 126.64t^5 + 304.67t^4 - 612.86t^3 + 1502.8t^2 \quad (3)$$

These curves may represent the occurrence in the influent wastewater of a treatment step of three compounds whose characteristics are reported above.

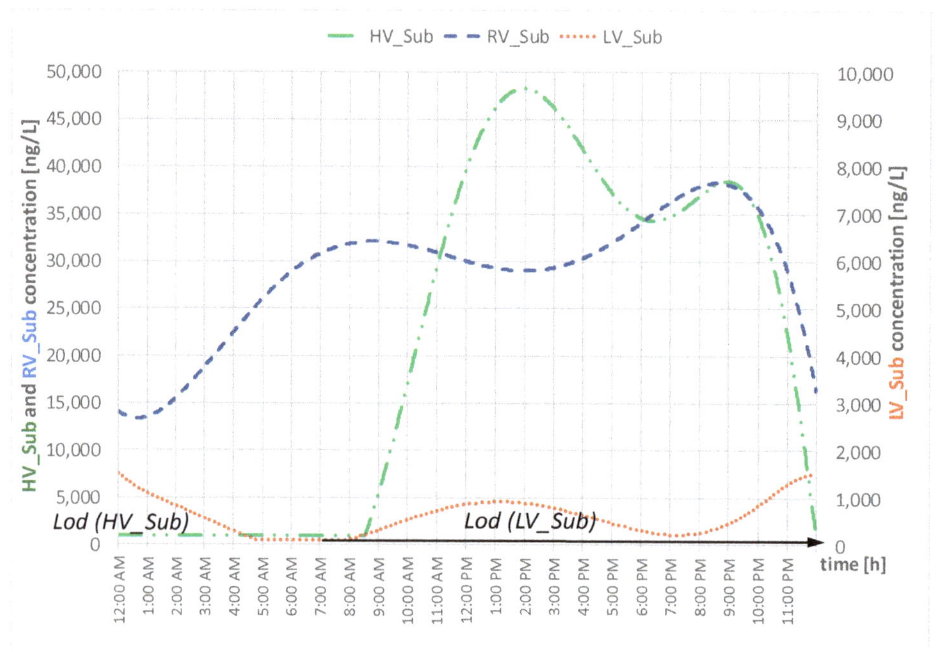

Figure 1. Concentrations versus time for the three key compounds considered in the study. Note that the Y-axis for the low variability substance (LV_Sub) is on the right and the Y-axis for the high variability substance (HV_Sub) and the random variability substance (RV_Sub) is on the left.

2.2. Flow Rate Curves Versus Time

It was assumed that the flow rate refers to the wastewater generated by a small catchment area (around 3500 inhabitants characterized by an individual water consumption of 200 L/(inhabitant day)) or a medium-large hospital (characterized by around 900 beds with a patient water consumption of 700 L/(patient day), according to literature [21]).

The selection of this size of wastewater source (small urban settlement or medium-large hospital) was in order to obtain more frequent and enhanced variations with regard to a larger urban settlement, as clearly shown by data provided in literature [12,22]. The flow rate referring to the whole day Q_{daily} is 634.5 m^3/d. Based on literature studies on curves of flow rate versus time (day) in settlement/hospital of this size [12,20,22], 24 values of flow rate were set (Table S1) and by software MATLAB R2018b a nonlinear regression was carried out leading to Equation (4) (Q is in m^3/h and time t in min). It is reported in Figure 2.

$$Q(t) = +0.01t^{11} - 0.11t^{10} + 0.78t^9 - 3.20t^8 + 7.12t^7 - 7.77t^6 + 5.10t^5 + 16.19t^4 \quad (4)$$

The wastewater volume flowing as a function of the time $V(t)$ is obtained by the integration of Equation (4):

$$V(t) = \int_{t=0}^{1440} Q(t)dt = 0.0001t^{12} - 0.01\,t^{11} + 0.078\,t^{10} - 0.35\,t^9 + 0.89\,t^8 - 1.11\,t^7 + 0.85\,t^6 + 3.24\,t^5 \quad (5)$$

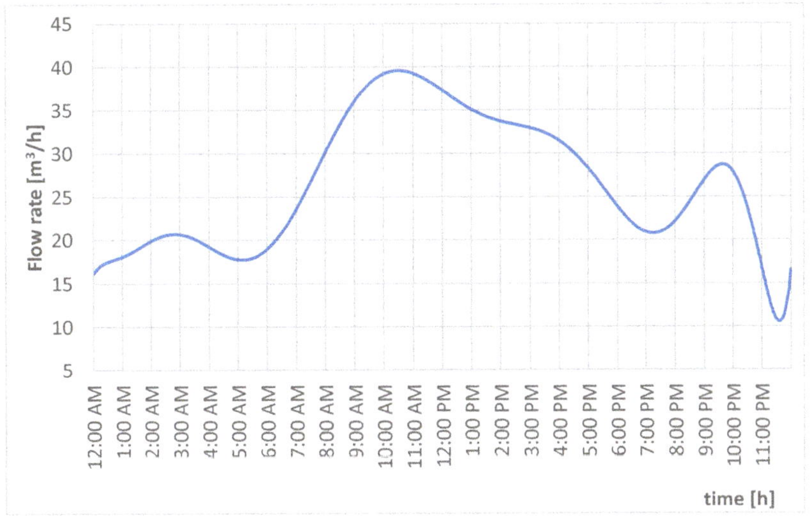

Figure 2. Flow rate versus time for the case study considered.

2.3. *The Sampling Modes Adopted and Compared*

The sampling modes compared in this study are those defined in Table 1:

Table 1. Description of the sampling modes adopted and compared in this study for average concentrations of the different compound.

Sampling	Description	Water Volume Sampled	Sampling Time, (Number of Samples)
Grab	The sampling consists of instantaneous (grab) wastewater withdrawal(s). The monitoring may include either one grab sample or a number of grab samples. The sampling time is defined by the investigation (monitoring protocol).	The requested wastewater volume for analysis	8 a.m. (1) 8 a.m. + 5 p.m. (2) 8 a.m. + 12 p.m. + 5 p.m. (3) 8 a.m. + 12 p.m. + 4 p.m. + 11 p.m. (4)
24-h time proportional composite	The sampling is performed at constant time intervals. It is the most common sampling mode. This is also called constant time, constant volume (CTCV)	A constant volume V_{sample} taken at each sampling instant	Every hour (24) Every 2 h (12) Every 4 h (6) Every 8 h (3)
24-h flow proportional composite	The sampling is performed at constant time intervals. The volume of wastewater taken is proportional to the flow rate flowing at each instant of sampling. This is also called constant time, variable volume (CTVV)	A linear interpolation curve is defined between the minimum and maximum wastewater flow and wastewater sampled over the whole observed range of variability of the wastewater flow (see Figure 3)	Every hour (24) Every 2 h (12) Every 4 h (6) Every 8 h (3)
24-h volume proportional composite	The sampling takes the same wastewater volume at variable time intervals, after a defined volume of wastewater has passed the sampling point. This is also called constant volume, variable time (CVVT)	A constant volume V_{sample} is taken at each defined sampling time	Frequency: Three times a day (3) Six times a day (6) Twelve times a day (12) Twenty-four times a day (24)

With regard to the flow proportional sampling mode, in order to define the direct proportionality curve between wastewater to be sampled and the flowing wastewater flow rate, the expected range of variability of the flow rate has to be known. In the case study, the observed range varied between 17.3 m^3/h and 38 m^3/h but, for the sake of caution, it was supposed that it might vary between 10 m^3/h and 50 m^3/h. It was then supposed that in the case of a flow rate of 10 m^3/h, the volume to sample would be equal to 20 mL, and in the case of 50 m^3/h, the volume to sample would be 100 mL, resulting in the linear relationship between volume to sample (y) and flow rate (x) y = 2x (Figure 3). Other direct proportional curves could be assumed for different cases.

In order to complete the analysis and the comparison among the available sampling strategies, the Supplementary Material contains Figures S1–S4 showing some details of the different sampling modes. Each graph remarks on the number and volume of samples withdrawn and the instant at which wastewater samples are taken in order to have all the information necessary to obtain the average concentration of the compound under study according to the adopted sampling approach. The flow rate curve versus time is also drawn in order to remark how variations in the flow rate may affect the evaluation of the micropollutant average concentration.

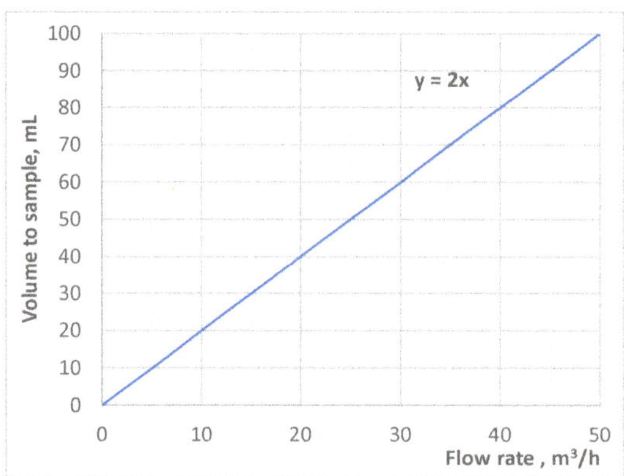

Figure 3. Relationship (direct proportionality) between volume to sample and flow rate for the flow proportional sampling mode.

2.4. Daily Average Concentration Evaluation

The ideal (true) obtainable concentration c_{ideal} for each compound was evaluated by means of Equation (6):

$$c_{ideal} = \frac{\sum_{i=1}^{1440} c_i\, Q_i}{\sum_{i=1}^{1440} Q_i} \tag{6}$$

where c_i is the concentration (ng/L) at minute i (in total 60 × 24 min = 1440 min) and Q_i is the flow rate (L/min) at the same minute i. Note that Q_i is numerically equal to the volume flowing during the minute i (V_i). The concentration value can be considered an accurate value (on a minute measurement basis) of the concentration of the compound. A shorter time interval could also be assumed, for instance the second, and in this case i varies up to 86,400.

The average concentrations of the key compounds were evaluated by assuming the different sampling modes. Note that, with regard to Figure 1, for the substances HV_Sub and LV_Sub, during the night their concentrations decrease below the corresponding Lod (according to the adopted analytical methods, but this issue is beyond the current study). For the sake of caution, it was assumed that the concentration was equal to the Lod (respectively 1000 ng/L and 100 ng/L). In addition, their corresponding limits of quantification (Loq) were assumed equal to 2500 ng/L and 250 ng/L: when their concentration was below the corresponding Loq, it was set equal to 0.5 Loq, according to [24].

In the case of grab sampling, the daily average concentration $\bar{c}_{grab}$ (ng/L) of a compound in the wastewater is based on the number n of the water grab samples withdrawn (Equation (7)). They were assumed to be 1, 2, 3 or 4 (as described in Table 1):

$$\bar{c}_{grab} = \frac{\sum_{i=1}^{n} c_i}{n}, \quad n = 1,\ 2,\ 3,\ 4 \tag{7}$$

where c_i is the concentration of the key compound in sample i in ng/L.

In the case of 24-h time proportional composite sampling, the daily average concentration of the key substance $\bar{c}_{time\ prop}$ was evaluated according to Equation (8):

$$\bar{c}_{time\ prop} = \frac{\sum_{i=1}^{k} c_i V_{sample}}{k\ V_{sample}} = V_{sample} \frac{\sum_{i=1}^{k} c_i}{k\ V_{sample}} = \frac{\sum_{i=1}^{k} c_i}{k},\ k = 24,\ 12,\ 6,\ 3 \quad (8)$$

where c_i is the concentration (ng/L) of the key compound in sample i, V_{sample} is the wastewater volume sampled (mL) at each withdrawal (always the same) and k is the number of samples taken according to the defined monitoring protocol (Table 1).

In the case of 24-h flow proportional composite sampling, the daily average concentration of the key substance $\bar{c}_{flow\ prop}$ (ng/L) was evaluated according to Equation (9):

$$\bar{c}_{flow\ prop} = \frac{\sum_{i=1}^{k} c_i\ \alpha\ Q_i}{\sum_{i=1}^{k} \alpha\ Q_i},\ k = 24,\ 12,\ 6,\ 3 \quad (9)$$

where c_i is the concentration of the key compound in sample i, in ng/L, $\alpha\ Q_i$ is the withdrawn wastewater volume (mL), α being the coefficient of direct proportionality (equal to 2) between the flow rate Q_i flowing at the sampling point at that instant and the volume to be sampled (see graph in Figure 2).

In the case of 24-h volume proportional composite sampling, the daily average concentration of the key substance $\bar{c}_{volume\ prop}$ (ng/L) was evaluated according to Equation (10):

$$\bar{c}_{volume\ prop} = \frac{\sum_{i=1}^{k} c_i\ V_{sample}}{k\ V_{sample}} = V_{sample} \frac{\sum_{i=1}^{k} c_i}{k\ V_{sample}} = \frac{\sum_{i=1}^{k} c_i}{k},\ k = 24,\ 12,\ 6,\ 3 \quad (10)$$

where c_i is the concentration of the key compound in sample i, in ng/L, and V_{sample} the wastewater volume (mL) sampled exactly after that the defined fraction $\frac{1}{k}$ of the daily volume of wastewater produced (V_{daily}) is flowed. Note that numerically, V_{daily} corresponds to Q_{daily}.

For the sake of clarity, it is here reported the sequence of steps necessary to obtain the average concentrations resulting from applying the different sampling modes described in Table 1. For grab sampling, time proportional and flow proportional composite sampling modes, the steps are:

1. definition of the sampling times according to Table 1;
2. calculation of the values of concentrations at each sampling time defined in the last column of Table 1 for the representative compound under study by the corresponding curve (Equations (1)–(3));
3. evaluation of the average daily concentration by applying the equation corresponding to the selected sampling mode (Equations (7)–(9)).

For the volume proportional composite sampling mode, the steps are:

1. definition of the frequency of sampling (k samples), according to the last column of Table 1 and the wastewater volume V_{vp} ($=V_{daily}/k$) which has to flow before collecting a water sample;
2. evaluation of the k sampling instants t_n, by means of the $V(t)$ curve (Equation (5)) posing $V(t_n) = n\ V_{vp}$ with $n = 1, \ldots, k$;
3. calculation of the values of concentrations at each sampling time t_n by the corresponding curve (Equations (1)–(3));
4. evaluation of the average daily concentration by applying Equation (10).

2.5. Mass Load Evaluation

The daily mass load ML (ng/d) of each substance can be evaluated as the product of the average concentration of the compound of interest $\bar{c}$ (ng/L) according to the different sampling modes

(Equations (7)–(9)) and the daily flow rate Q_{daily} (L/d). It is clear that this is directly proportional to the average concentrations through the daily flow rate (=634.5 m^3/d).

$$ML = \bar{c}\, Q_{daily} \tag{11}$$

2.6. Removal Efficiency Evaluation of a Micropollutant: Considerations and Remarks

As discussed in [25], with regard to a generic wastewater treatment step (Figure 4), the percentage efficiency μ in removing a specific contaminant j is defined on the basis of the mass loading (corresponding to the product: concentration × flow rate) in its influent (stream number 1) and effluents (stream numbers 2 and 3) at a set time interval, in accordance with Equation (12):

$$\mu_{total,\, j} = \frac{c_{1,j}\, Q_1 - \left(c_{2,j}\, Q_2 + c_{3,j}\, Q_3\right)}{c_{1,j}\, Q_1} \times 100 \tag{12}$$

As reported in the caption of Figure 4, the step could produce two different effluents (as in a conventional activated sludge system or in a membrane bioreactor: the clarified effluent or the permeate and the excess sludge). Quite often, the equation used for the evaluation of removal efficiency in an activated sludge system does not consider the occurrence of the (micro)pollutant in the excess sludge (this assumes $c_{3,j} = 0$) and, as reported in [25], this leads to an "apparent" removal efficiency, generally higher than the total removal efficiency.

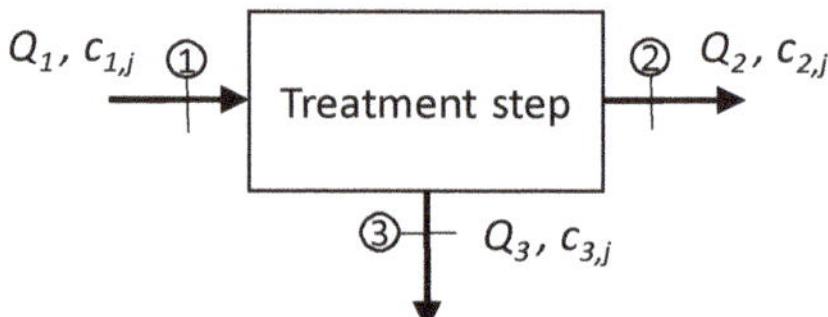

Figure 4. Representation of a generic wastewater treatment step, for instance an activated sludge system with the two effluents: a liquid phase (the clarified effluent, stream number 2) and the solid phase that is the excess sludge (stream number 3). In the case of a treatment step with only one effluent stream, stream number 3 does not appear.

In this study we have evaluated removal efficiency in the case of a treatment step with only one effluent stream (namely a polishing treatment by constructed wetlands, lagoons, and rapid filtration). Moreover, the time interval assumed for its evaluation is the day, hence the micropollutant concentrations c_1 and c_2 (referring to the influent and the effluent) considered are the daily average concentrations obtained by following the different sampling modes described in Table 1, $Q_1 = Q_2$, and they are numerically equal to V_{daily}.

The ideal removal efficiency μ_{ideal} was evaluated by means of Equation (13):

$$\mu_{ideal} = \frac{\sum_{i=1}^{1440} c_{1,i}\, Q_{1,i} - \sum_{i=1}^{1440} c_{2,i}\, Q_{2,i}}{\sum_{i=1}^{1440} c_{1,i}\, Q_{1,i}} \tag{13}$$

where $c_{1,i}$ and $c_{2,i}$ are the micropollutant concentrations (ng/L) at minute i in the influent and effluent respectively, $Q_{1,i}$ and $Q_{2,i}$ are the flow rates (L/min) at minute i in the influent and effluent ($Q_{1,i} = Q_{2,i}$).

Case Study for the Evaluation of Removal Efficiency

The analysis of the removal efficiency evaluation refers to the data reported in Figure 5, which represents the profile of a randomly variable compound, as described in Section 2.1 for the influent and effluent of a small wastewater treatment plant, characterized by a hydraulic retention time (HRT) of 12 h.

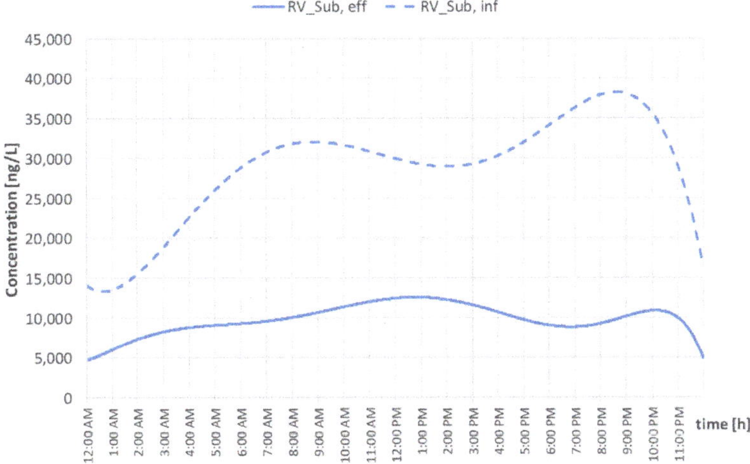

Figure 5. Occurrence of the same compound in a small wastewater treatment plant influent (dashed line) and effluent (continuous line).

The correlation between the concentration of the key compound and time (min) in the influent corresponds to Equation (2) and for the effluent, to Equation (14). This curve is obtained following the same procedure adopted for Equations (1)–(3) and it is based on the 24 raw data compiled in Table S1:

$$c_{RV\_Sub,\,eff}\,(t) = +0.09t^7 - 2.59t^6 + 34.09t^5 - 206.33t^4 - 408.29t^3 - 1082.9t^2 + 4736.7t^1 \quad (14)$$

where t is in minutes and $c_{RV\_Sub,\,eff}$ in ng/L.

The removal efficiency for the key compound was evaluated according to the different sampling modes defined in Table 2.

Table 2. Description of the sampling modes adopted and compared in this study for the removal efficiency evaluation of RV_Sub.

Sampling	Sampling Time for Influent and Effluent (Number of Samples)	Some Remarks and Number of Estimated Values of Removal Efficiencies in Brackets
Grab	Every hour (24), hydraulic retention time (HRT) not considered Every hour (24), HRT considered	Removal evaluated each hour (24 values)
	8 a.m.; 5 p.m. (2) HRT not considered 8 a.m.; 5 p.m. (2) HRT not considered 8 a.m.; 12 p.m.; 5 p.m. (3) HRT not considered 8 a.m.; 12 p.m.; 5 p.m. (3) HRT considered 8 a.m.; 12 p.m.; 4 p.m.; 11 p.m. (4) HRT not considered 8 a.m.; 12 p.m.; 4 p.m.; 11 p.m. (4) HRT considered	Removal based on average values for influent and effluent (one value)
Time proportional	24-h time proportional composite sample, time interval between two consecutive withdrawals equal to 1 h (1)	(One value)
Flow proportional	24-h flow proportional composite sample, time interval between two consecutive withdrawals equal to 1 h (1)	(One value)
Volume proportional	24-h volume proportional composite sample. Twenty-four samples a day mixed for the composite sample as reported in Table 1 (1)	(One value)

In addition, the removal efficiency of RV_Sub was also estimated, assuming that concentrations were known with a frequency equal to 1 min. This is considered the "ideally obtainable" value of removal efficiency. The collection of this amount of concentrations for many micropollutants is completely unrealistic, due to the high costs and time requested for their analytical determination.

3. Results

3.1. Average Concentration of the Key Compounds

The ideally obtainable daily average concentrations of the three representative compounds were found by applying Equation (6) and are reported in Table 3.

Table 3. Ideal average concentrations c_{ideal} for the three substances and corresponding standard deviation (SD) (c_{ideal} ± SD).

HV_Sub, ng/L	LV_Sub, ng/L	RV_Sub, ng/L
24,561 ± 18,305	586 ± 377	29,609 ± 6674

These values are compared here with the average concentrations resulting from applying the different sampling modes described in Table 1, according to the procedure described in Section 2.4. Details of the application of this procedure is reported in Tables S2–S4 with regard only to RV_Sub. For all the substances, the evaluated average concentrations are here reported in tables: Table 4 refers to the case of a different number of grab samples, Table 5 to 24-h time proportional composite sampling, Table 6 to flow proportional composite samples and finally, Table 7 to volume proportional composite samples.

Table 4. Average concentrations of the three substances in the case of grab samples (with the different number of samples collected).

Number (#) of Grab Samples	HV_Sub, ng/L	LV_Sub, ng/L	RV_Sub, ng/L
1	1000	112	31,852
2	19,041	287	31,954
3	26,014	478	31,301
4	26,117	724	30,263

Table 5. Average concentrations of the three substances in the case of time proportional sampling (with the different number of samples collected).

Interval (h), (#of Samples)	HV_Sub, ng/L	LV_Sub, ng/L	RV_Sub, ng/L
1 (24)	21,751	590	28,664
2 (12)	21,518	595	28,472
4 (6)	20,270	608	27,799
8 (3)	14,535	750	25,409

Table 6. Average concentrations of the three substances in the case of flow proportional sampling (with the different number of samples collected).

Interval (h), (#of Samples)	HV_Sub, ng/L	LV_Sub, ng/L	RV_Sub, ng/L
1 (24)	24,477	590	29,525
2 (12)	24,412	596	29,443
4 (6)	23,550	581	29,000
8 (3)	17,406	612	27,543

Table 7. Average concentrations of the three substances in the case of volume proportional sampling.

Frequency (#/d)	HV_Sub, ng/L	LV_Sub, ng/L	RV_Sub, ng/L
3	18,848	888	25,948
6	22,359	644	28,702
12	23,867	602	29,314
24	24,365	590	29,541

It emerges that for all three substances, average concentrations resulting from the grab sampling mode present the widest ranges of variability, whereas the 24-h flow proportional composite sampling show the smallest ranges of variability. Moreover, one grab sample may lead to an enhanced underestimation or overestimation, depending on the time of sampling and the concentration profile. In the case of a substance with a "flat" curve of concentrations versus time, a grab sample could be considered representative of the "average" daily concentration whatever time it is taken. But in all the other situations, a grab sample should be avoided.

An increment in the frequency of withdrawal for the composite sampling mode always leads to an average concentration measurement, which is closer to the ideal value, whatever the concentration profile.

With regard to the HV_Sub average concentrations reported in Tables 4–7, it emerges that the lowest value is 1000 ng/L, and the highest is 26,117 ng/L found with the grab sampling mode. This is due to the fact that this substance presents very low concentrations during the night (between 12:00 a.m. and 9:00 a.m. it was below its limit of detection (Lod) and for the sake of caution, was assumed to be equal to its Lod value) and the lowest value corresponds to one grab sample taken at 8:00 a.m. and the highest to four grab samples taken at 8:00 a.m., 12:00 p.m., 4:00 p.m. and 11:00 p.m., with only one sample collected in the interval in which concentrations are very low, assumed equal to the corresponding Lod (1000 ng/L).

With regard to LV_Sub, the lowest average concentration was found with one grab sample (112 ng/L) and the highest with the 24-h volume proportional sample, with samples taken three times a day (818 ng/L).

Finally, referring to RV_Sub, the highest average concentration was found with the grab sample taken at 8:00 a.m. and the lowest average concentration with the 24-h composite sampling mode (three samples taken every eight hours). It is important to observe that the highest value does not correspond to the maximum concentration of the RV_Sub profile of concentration: 38,298 ng/L occurring at 8:35 p.m.

For each of the three substances, the percentage deviation ($=\frac{c - c_{ideal}}{c_{ideal}} \times 100$) between the ideal average concentration c_{ideal} (see Table 3) and the "measured" average concentrations obtained following a specific sampling mode are reported in the three "target" diagrams in Figure 6. The circumferences refer to percentage deviations (1%, 10%, 40% and 100%) on a logarithmic scale. Full symbols represent situations in which the average measured concentration is higher than the corresponding ideal concentration (overestimation) and empty symbols to situations in which the average measured concentration is lower than the corresponding ideal concentration (underestimation).

It emerges that for all three compounds, the sampling mode and frequency which lead to the best estimation of the average concentration are always the 24-h flow proportional composite sampling with samples taken every hour and the 24-h volume proportional composite sampling with 24 samples per day. Moreover, the sampling mode with the smallest deviation is flow proportional: the deviation always remains below 10% with only one exception (HV_Sub with samples taken every eight hours).

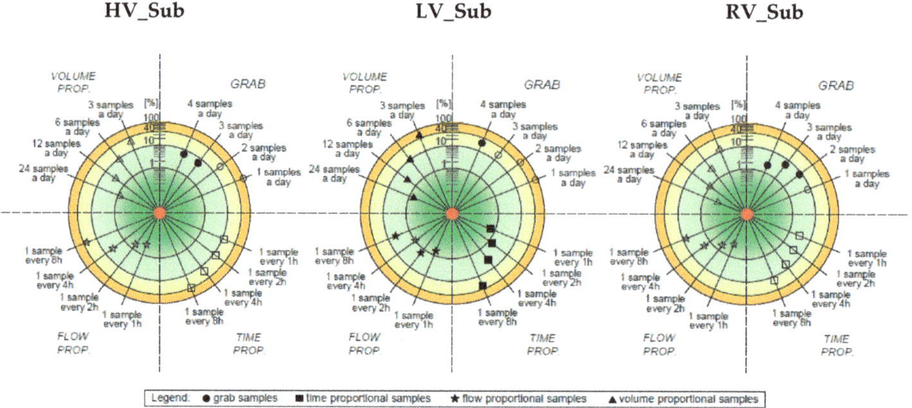

Figure 6. Percentage deviations between the ideal concentration of each substance (red dot) and the measured average concentrations found following the different sampling modes, defined in Table 1. Circumferences in the three graphs refer to the different values of percentage deviation on a log scale. Full symbols correspond to an overestimation and empty symbols to an underestimation.

It is interesting to observe that the "measured" average concentration is only overestimated (full symbol) for LV_Sub, whereas for HV_Sub and RV_Sub measured average concentrations are underestimated (empty symbols), with just a few exceptions. This fact can be explained by the different concentration profiles versus time of the compounds. Figure 1 shows that LV_Sub is the only compound with night concentrations even higher than diurnal ones and, in the case of time and volume proportional composite samplings (which do not consider the weight of the flow rate) this leads to an overestimation. The ideal average concentration, as shown by the definition in equation 6, weights the concentration with the flow rate, which is lower during the night (Figure 2).

These considerations provide a good explanation as to why time proportional composite sampling could be a good mode for RV_Sub. For this substance, the range of percentage deviations is the smallest in comparison to the range of the other two compounds.

The analysis of the different average concentration values for the three compounds highlights that the selection of the sampling mode which is best suited to the aim of the monitoring campaign depends on the type of substance and on its expected concentration profiles, if known. It could be of interest to know the average concentration of the compound in order to design a treatment train capable of removing it. It could also be of interest to know the highest concentration during the day in case an environmental risk assessment should be carried out (in this case, the European Guidelines [26–28] suggest taking the maximum concentration of a compound in order to consider the worst-case scenario). In fact, if the substance has very low concentrations during the night or in well-known daytime intervals, monitoring planning could avoid this period.

3.2. Mass Load Evaluated for Each Substance

The ideal mass loading of each substance was evaluated by Equation (11) and is reported in Table 8. As highlighted in Section 2.5, the percentage deviations with respect to the ideal value of the mass loading of each substance is the same as those found for the average concentrations with regard to the same sampling mode and (obviously) substance.

Table 8. Evaluation of the mass load for the three substances.

HV_Sub, g/d	LV_Sub, g/d	RV_Sub, g/d
15.6	0.37	18.8

3.3. Average Removal Efficiency for RV-Sub

The ideally obtainable daily removal efficiency was obtained by applying Equation (13) and is equal to 67.8%. On the basis of the average daily concentrations in the influent and effluent obtained by the different sampling modes (Table 2), the corresponding removal efficiencies were evaluated.

In the first case, a grab sample mode is followed; the flow rate at the entrance and exit of the treatment step is assumed to be the same and samples are taken at the same time (HRT of the treatment step is not considered). The removal efficiency based on only one grab sample during a day and varies between 53% and 76% depending on the sampling time. If samples are taken considering the HRT of the plant (12 h), the removal efficiency varies between 7% and 84%, always depending on the sampling times at the two points. Figure 7 reports the values in both scenarios.

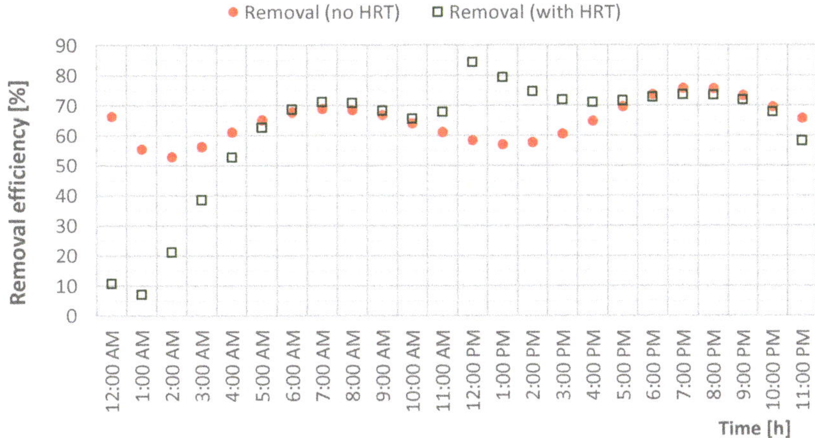

Figure 7. Evaluated removal efficiency of the RV_Sub on the basis of single grab samples taken at the influent and effluent at the same time (colored circle) and considering the hydraulic retention time (HRT) of the treatment step (void square).

Table 9 reports the RV_Sub removal efficiencies in the case of a different numbers (2, 3, 4) of grab samples taken in the influent and effluent, at the same time (case 1) and considering the HRT of the plant (case 2). It is important to underline that the removal efficiency is evaluated on the basis of the average values in the influent and the effluent of the n grab samples taken as remarked in the last column of Table 2.

Table 9. Removal efficiency of RV_Sub in the case of grab samples in different scenarios.

Number of Grab Samples	Case 1: HRT not Considered	Case 2: HRT Considered
2	68.9	71.2
3	65.9	75.3
4	64.3	71.1

It emerges that when HRT is considered, the removal efficiency is always higher than when it is neglected, and it is also higher than the ideally obtainable removal efficiency (equal to 67.8%).

In the case of 24-h time proportional composite sampling, the daily removal efficiency was equal to 65.8%; in the case of 24-h flow proportional composite sampling, the daily removal efficiency was equal to 67.8%, and in the case of 24-h volume proportional composite sampling, the efficiency was 65.4%, all of which are very close to the ideal removal efficiency (67.8%).

The target graph in Figure 8 reports and compares the percentage deviations ($=\frac{\mu - \mu_{ideal}}{\mu_{ideal}} \times 100$) between the ideal removal efficiency μ_{ideal} (67.8%, corresponding to the red circle in the center) and the values μ found following the different sampling modes.

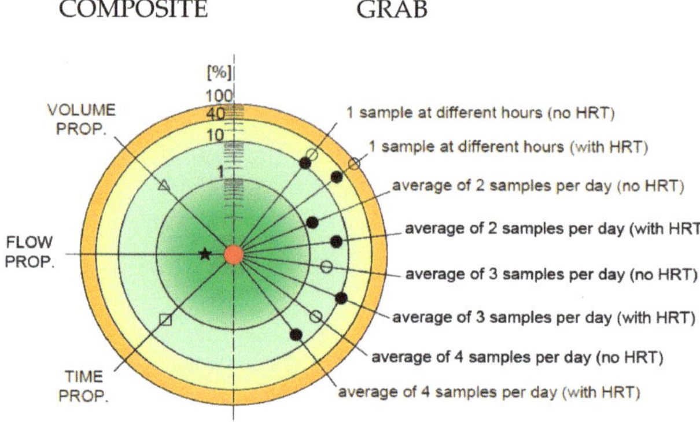

Figure 8. Percentage deviations between ideal removal efficiency for RV_Sub (red dot) and the evaluated removal efficiency found following the different sampling modes, defined in Table 1. Circumferences in the three graphs refer to different values of percentage deviation on a log scale. Full symbols correspond to an overestimation and empty symbols to an underestimation.

It emerges that, in the case of removal efficiency based on one grab sample, the ranges of percentage deviations vary between (−22%; +11.5%) without considering HRT and between (−89.6%; +24.2%) if HRT is considered. In case of more grab samples taken in a day, (considering or not considering the HRT), the percentage deviation remains between −5.0% and 11.1%. The flow proportional composite sampling mode leads to the most accurate evaluation (0.06%), compared to time proportional (−2.9%) and volume proportional (−3.5%) modes.

4. Discussion and Final Remarks

This study highlights the influence of the sampling mode on the collected measured concentrations of micropollutants which present different concentration profiles versus time (over the day). It also compares the removal efficiencies achieved in an ideal treatment step when influent and effluent concentrations are collected following different sampling modes. Unfortunately, this is not always reported and described in published papers, as highlighted by [1,5]. In particular in [1], a review dealing with the removal of pharmaceuticals from wastewater by different constructed wetlands, an analysis of the information regarding the adopted sampling modes allows the reader to "weigh up/assess" the reliability of the collected data presented. The most adopted mode in the case of monitoring campaigns regarding micropollutants in different water environments is that of 24-h time proportional composite sampling, whatever the micropollutant and its occurrence profile.

The three "ideal" substances considered in the current study are representative of three different cases and give some insights into the expected scenarios a researcher could find in investigating campaigns in terms of monitoring the occurrence and evaluating the removal of micropollutants. The analysis reported and discussed here provides some figures regarding the expected deviations with regard to ideal values, also called the 'true' concentrations and the "true" removal efficiency of a micropollutant. It was found that the flow proportional composite sampling mode leads to the best evaluation of the average concentration of a micropollutant (whatever the concentration profile is) and also of its removal efficiency. It is followed by the volume proportional composite sampling mode, and then by the time proportional one. The grab sample can be adopted if the number of collected samples is able to catch the main (expected) variations of concentrations over the day, in particular when the concentration curve versus time is flat, or when the aim of the monitoring campaign is to find the maximum concentration during the day in case of environmental risk assessment and it is known when it may occur. As most of the micropollutants are unregulated compounds, guidelines for sampling campaigns dealing with them are not available.

To complete the discussion on reliability of collected (measured data) it is important to spend some words on the issue of the uncertainties associated with the direct measurements of concentrations in the water environment. In the current study, it was found that the average evaluated concentrations obtained by applying Equations (1)–(3) and (12) lead to an uncertainty varying in the range between <1% and 30% for 24-h flow proportional composite sampling, between <1% and 40% for 24-h time proportional composite sampling, between <1% and up to 51% for 24-h volume proportional composite sampling and even up to 95% in case of one grab sample in a day.

These values are in agreement with other studies which found uncertainties varying from 10% in the case of 24-h flow proportional composite sampling [17,29] to 25% (even 100%) if time proportional composite sampling is adopted [30]. Regarding uncertainties associated with chemical analysis, literature studies found that they are lower than those for sampling; they may vary between 4% and 16% [31]. Finally, uncertainties in flow rate measurement may vary between 6% according to [17] to 20% according to [32].

These considerations underline the importance of properly defining a sampling mode in order to provide highly reliable data regarding the occurrence and also removal of micropollutants from wastewater.

Supplementary Materials: The following are available online at http://www.mdpi.com/2073-4441/11/6/1152/s1, Table S1: Concentrations of the three representative compounds and values of flow rates used for defining the corresponding profile of concentrations and flow rate over the day (Figures 1, 2 and 5 in the manuscript); Table S2–S4: Evaluation of the average concentrations of the three representative compounds following the different sampling modes; Figure S1: Flow rate profile (dashes) and withdrawn volume (full circles) for each grab sample. Note the volume is always the same at the defined instants of time in case of four grab samples (i.e., 8:00 a.m.; 12:00 p.m.; 5:00 p.m. and 11:00 p.m.); Figure S2: Flow rate profile (dashes) and volume withdrawn (full circles) for the 12 water samples. Also in this case, the sample volume is constant. Samples are taken every 2 h; Figure S3: Flow rate profile (dashes) and volume withdrawn (full circles) for the 12 water samples. The volume taken for the different samples is proportional to the flow rate at the sampling time. Samples are taken every 2 h; Figure S4. Flow rate (dashes) profile and volume withdrawn (full circles) for the 12 water samples. The volume taken for the different samples is constant. Samples are taken when $\frac{1}{12}Q_{daily}$ is passed at the sampling point.

Author Contributions: For this research article, both authors contributed extensively to the work. P.V. supervised the research and defined the methodologies; A.G. analyzed the data, P.V: and A.G. contributed to the writing and the review of the paper.

Funding: This research received University of Ferrara funding.

Conflicts of Interest: The authors declare no conflict of interest.

References

1. Verlicchi, P.; Zambello, E. How Efficient are Constructed Wetlands in Removing Pharmaceuticals from Untreated and Treated Urban Wastewaters? A Review. *Sci. Total Environ.* **2014**, *470*, 1281–1306. [CrossRef] [PubMed]
2. Weissbrodt, D.; Kovalova, L.; Ort, C.; Pazhepurackel, V.; Moser, R.; Hollender, J.; Siegrist, H.; Mcardell, C.S. Mass flows of X-ray contrast media and Cytostatics in hospital wastewater. *Environ. Sci. Technol.* **2009**, *43*, 4810–4817. [CrossRef] [PubMed]
3. Verlicchi, P. Pharmaceutical Concentrations and Loads in Hospital Effluents: Is a Predictive Model or Direct Measurement the Most Accurate Approach? In *Hospital Wastewater—Characteristics, Management, Treatment and Environmental Risks*; Springer: Cham, Switzerland, 2018; pp. 101–134. [CrossRef]
4. Kovalova, L.; Siegrist, H.; Singer, H.; Wittmer, A.; McArdell, C.S. Hospital wastewater treatment by membrane bioreactor: Performance and efficiency for organic micropollutant elimination. *Environ. Sci. Technol.* **2012**, *46*, 1536–1545. [CrossRef] [PubMed]

5. Ort, C.; Lawrence, M.G.; Reungoat, J.; Mueller, J.F. Sampling for PPCPs in wastewater systems: Comparison of different sampling modes and optimization strategies. *Environ. Sci. Technol.* **2010**, *44*, 6289–6296. [CrossRef] [PubMed]
6. Ort, C.; Lawrence, M.G.; Rieckermann, J.; Joss, A. Sampling of pharmaceuticals and personal care products (PPCPs) and illicit drugs in wastewater systems: Are your conclusions valid? A critical review. *Environ. Sci. Technol.* **2010**, *44*, 6024–6035. [CrossRef] [PubMed]
7. Peña-Guzmán, C.; Ulloa-Sánchez, S.; Mora, K.; Helena-Bustos, R.; Lopez-Barrera, E.; Alvarez, J.; Rodriguez-Pinzón, M. Emerging pollutants in the urban water cycle in Latin America: A review of the current literature. *J. Environ. Manag.* **2019**, *237*, 408–423. [CrossRef] [PubMed]
8. Fekadu, S.; Alemayehu, E.; Dewil, R.; Van der Bruggen, B. Pharmaceuticals in freshwater aquatic environments: A comparison of the African and European challenge. *Sci. Total Environ.* **2019**, *654*, 324–337. [CrossRef]
9. Yap, H.C.; Pang, Y.L.; Lim, S.; Abdullah, A.Z.; Ong, H.C.; Wu, C. A comprehensive review on state-of-the-art photo-, sono-, and sonophotocatalytic treatments to degrade emerging contaminants. *Int. J. Environ. Sci. Technol.* **2019**, *16*, 601–628. [CrossRef]
10. Cattaneo, S.; Marciano, F.; Masotti, L.; Vecchiato, G.; Verlicchi, P.; Zaffaroni, C. Improvement in the Removal of Micropollutants at Porto Marghera Industrial Wastewaters Treatment Plant by MBR Technology. *Water Sci. Technol.* **2008**, *58*, 1789–1796. [CrossRef]
11. Paíga, P.; Correia, M.; Fernandes, M.J.; Silva, A.; Carvalho, M.; Vieira, J.; Jorge, S.; Silva, J.G.; Freire, C.; Delerue-Matos, C. Assessment of 83 pharmaceuticals in WWTP influent and effluent samples by UHPLC-MS/MS: Hourly variation. *Sci. Total. Environ.* **2019**, *648*, 582–600. [CrossRef]
12. Duong, H.; Pham, N.; Nguyen, H.; Hoang, T.; Pham, H.; CaPham, V.; Berg, M.; Giger, W.; Alder, A. Occurrence, fate and antibiotic resistance of fluoroquinolone antibacterials in hospital wastewaters in Hanoi, Vietnam. *Chemosphere* **2008**, *72*, 968–973. [CrossRef] [PubMed]
13. Nelson, E.D.; Do, H.; Lewis, R.S.; Carr, S.A. Diurnal variability of pharmaceutical, personal care product, estrogen and alkylphenol concentrations in effluent from a tertiary wastewater treatment facility. *Environ. Sci. Technol.* **2011**, *45*, 1228–1234. [CrossRef] [PubMed]
14. Burns, E.E.; Carter, L.J.; Kolpin, D.W.; Thomas-Oates, J.; Boxall, A.B.A. Temporal and spatial variation in pharmaceutical concentrations in an urban river system. *Water Res.* **2018**, *137*, 72–85. [CrossRef] [PubMed]
15. Barbosa, M.O.; Ribeiro, A.R.; Ratola, N.; Hain, E.; Homem, V.; Pereira, M.F.R.; Blaney, L.; Silva, A.M.T. Spatial and seasonal occurrence of micropollutants in four Portuguese rivers and a case study for fluorescence excitation-emission matrices. *Sci. Total Environ.* **2018**, *644*, 1128–1140. [CrossRef] [PubMed]
16. Ort, C.; Gujer, W. Sampling for representative micropollutant loads in sewer systems. *Water Sci. Technol.* **2006**, *54*, 169–176. [CrossRef] [PubMed]
17. Ort, C.; Lawrence, M.G.; Reungoat, J.; Eaglesham, G.; Carter, S.; Keller, J. Determining the fraction of pharmaceutical residues in wastewater originating from a hospital. *Water Res.* **2010**, *44*, 605–615. [CrossRef] [PubMed]
18. Kummerer, K.; Helmers, E. Hospital effluents as a source of gadolinium in the aquatic environment. *Environ. Sci. Technol.* **2000**, *34*, 573–577. [CrossRef]
19. Khan, S.; Ongerth, J. Occurrence and removal of pharmaceuticals at an Australian sewage treatment plant. *Water* **2005**, *32*, 80–85.
20. Verlicchi, P.; Galletti, A.; Al Aukidy, M. Hospital Wastewaters: Quali-quantitative Characterization and Strategies for Their Treatment and Disposal. In *Wastewater Reuse and Management*; Springer Science & Business Media: New York, NY, USA, 2013; pp. 225–251.
21. Verlicchi, P.; Galletti, A.; Petrovic, M.; Barcelo, D. Hospital Effluents as a Source of Emerging Pollutants: An Overview of Micropollutants and Sustainable Treatment Options. *J. Hydrol.* **2010**, *389*, 416–428. [CrossRef]
22. Boillot, C.; Bazin, C.; Tissot-Guerraz, F.; Droguet, J.; Perraud, M.; Cetre, J.C.; Trepo, D.; Perrodin, Y. Daily physicochemical, microbiological and ecotoxicological fluctuations of a hospital effluent according to technical and care activities. *Sci. Total Environ.* **2008**, *403*, 113–129. [CrossRef]
23. Hong, Y.; Sharma, V.K.; Chiang, P-C.; Kim, H. Fast-target analysis and hourly variation of 60 pharmaceuticals in wastewater using UPLC-high resolution mass spectrometry. *Arch. Environ. Contam. Toxicol.* **2015**, *69*, 525–534. [CrossRef] [PubMed]
24. Armbruster, D.A.; Pry, T. Limit of Blank, Limit of Detection and Limit of Quantitation. *Clin. Biochem. Rev.* **2008**, *29*, S49. [PubMed]

25. Verlicchi, P.; Al Aukidy, M.; Zambello, E. Occurrence of Pharmaceutical Compounds in Urban Wastewater: Removal, Mass Load and Environmental Risk After a Secondary Treatment-A Review. *Sci. Total Environ.* **2012**, *429*, 123–155. [CrossRef] [PubMed]
26. European Community (EC). Technical Guidance Document on Risk Assessment in support of Commission Directive 93/67/EEC on Risk Assessment for new notified substances, Commission Regulation (EC) No 1488/94 on Risk Assessment for Existing Substances, and Directive 98/8/EC of the European Parliament and of the Council concerning the placing of biocidal products on the market, Parts I. European Communities, EUR 20418 EN/1. 2003. Available online: https://echa.europa.eu/documents/10162/16960216/tgdpart1_2ed_en.pdf (accessed on 31 May 2019).
27. European Community (EC). Technical Guidance Document on Risk Assessment in support of Commission Directive 93/67/EEC on Risk Assessment for new notified substances, Commission Regulation (EC) No 1488/94 on Risk Assessment for Existing Substances, and Directive 98/8/EC of the European Parliament and of the Council concerning the placing of biocidal products on the market, Parts II. European Communities, EUR 20418 EN/1. 2003. Available online: https://echa.europa.eu/documents/10162/16960216/tgdpart2_2ed_en.pdf (accessed on 31 May 2019).
28. European Community (EC). Technical Guidance Document on Risk Assessment in support of Commission Directive 93/67/EEC on Risk Assessment for new notified substances, Commission Regulation (EC) No 1488/94 on Risk Assessment for Existing Substances, and Directive 98/8/EC of the European Parliament and of the Council concerning the placing of biocidal products on the market, Parts IV. European Communities, EUR 20418 EN/1. 2003. Available online: https://echa.europa.eu/documents/10162/16960216/tgdpart4_2ed_en.pdf (accessed on 31 May 2019).
29. Jelic, A.; Fatone, F.; Di Fabio, S.; Petrovic, M.; Cecchi, F.; Barcelo, D. Tracing pharmaceuticals in a municipal plant for integrated wastewater and organic solid waste treatment. *Sci. Total Environ.* **2012**, *433*, 352–361. [CrossRef] [PubMed]
30. Verlicchi, P.; Zambello, E. Predicted and measured concentrations of pharmaceuticals in hospital effluents. Examination of the strengths and weaknesses of the two approaches through the analysis of a case study. *Sci. Total Environ.* **2016**, *565*, 82–94. [CrossRef] [PubMed]
31. Verlicchi, P.; Al Aukidy, M.; Jelic, A.; Petrović, M.; Barceló, D. Comparison of Measured and Predicted Concentrations of Selected Pharmaceuticals in Wastewater and Surface Water: A Case Study of a Catchment Area in the Po Valley (Italy). *Sci. Total Environ.* **2014**, *470*, 844–854. [CrossRef] [PubMed]
32. Lai, F.Y.; Ort, C.; Gartner, C.; Carter, S.; Prichard, J.; Kirkbride, P.; Bruno, R.; Hall, W.; Eaglesham, G.; Mueller, J.F. Refining the estimation of illicit drug consumptions from wastewater analysis: Co-analysis of prescription pharmaceuticals and uncertainty assessment. *Water Res.* **2011**, *45*, 4437–4448. [CrossRef]

Article

Spectrophotometric Detection of Glyphosate in Water by Complex Formation between Bis 5-Phenyldipyrrinate of Nickel (II) and Glyphosate

Aline Romero-Natale [1], Ilaria Palchetti [2], Mayra Avelar [3,4], Enrique González-Vergara [1], José Luis Garate-Morales [5] and Eduardo Torres [1,*]

1 Centro de Química, ICUAP, Benemérita Universidad Autónoma de Puebla, 14 Sur y Av. San Claudio, Col. San Manuel, Puebla 72570, Mexico; aline_natale@hotmail.com (A.R.-N.); enrique.gonzalez@correo.buap.mx (E.G.-V.)

2 Dipartimento di Chimica, Università Degli Studi di Firenze, Via della Lastruccia 3, 50019 Sesto Fiorentino (Fi), Italy; ilaria.palchetti@unifi.it

3 Departamento de Ingeniería Celular y Biocatálisis, Instituto de Biotecnología Universidad Nacional Autónoma de México, Av. Universidad 2001 Chamilpa, Cuernavaca 62210, Morelos, Mexico; mavelarf@gmail.com

4 Department of Biotechnology, Chemistry, and Pharmacy, University of Siena, Via A. Moro 2, 53100 Siena, Italy

5 Facultad de Ciencias Químicas, Benemérita Universidad Autónoma de Puebla, 14 Sur y Av. San Claudio, Col. San Manuel, Puebla 72570, Mexico; jose.garate@correo.buap.mx

* Correspondence: eduardo.torres@correo.buap.mx; Tel.: +52-222-2295500 (ext. 7273)

Received: 26 February 2019; Accepted: 3 April 2019; Published: 6 April 2019

Abstract: A spectrophotometric method for the determination of glyphosate based on the monitoring of a complex formation between bis 5-phenyldipyrrinate of nickel (II) and the herbicide was developed. The method showed a short response time (10 s), high selectivity (very low interference from other pesticides and salts), and high sensitivity (LOD 2.07×10^{-7} mol/L, LOQ 9.87×10^{-7} mol/L, and a K_d from 1.75×10^{-6} to 6.95×10^{-6} mol/L). The Job plot showed that complex formation occurs with a 1:1 stoichiometry. The method was successfully applied in potable, urban, groundwater, and residual-treated water samples, showing high precision (0.34–2.9%) and accuracy (87.20–119.04%). The structure of the complex was elucidated through theoretical studies demonstrating that the nickel in the bis 5-phenyldipyrrinate forms a distorted octahedral molecular geometry by expanding its coordination number through one bond with the nitrogen and another with the oxygen of the glyphosate' carboxyl group, at distances between 1.89–2.08 Å.

Keywords: environmental analysis; glyphosate; pesticides; phenyldipyrrinate; spectrophotometry; water pollution

---

1. Introduction

Glyphosate (N-(phosphonomethyl)glycine) (glyp) (Figure 1c) is the most intensively used herbicide worldwide because of its high effectiveness against annual grasses and aquatic weeds [1,2], with a global annual production estimated to be over 825,804 tons in 2014 [3]. Its use is allowed in agricultural, urban, and domestic activities [4], with the agricultural application being the most intensive one. The physical and chemical properties of glyp allow its distribution in the environmental compartments [5–7]. Furthermore, its chelating ability and the absorption constant in soil allows for its accumulation in several types of soils, mainly clays [8]; it is considered a stable compound in a pH range between 4 and 9 for hydrolysis and photolysis [9,10]. Glyp pollution in water, soil, and food samples are becoming a serious health concern [11–13]. In addition, its metabolite, aminomethylphosphonic acid (AMPA), also represents a potential danger to human and animal health [14,15]. Both compounds have been detected in groundwater and surface water in several countries [6,16,17]. As glyp has been labeled as a global pollutant [18,19], assessing its presence in several environmental matrices is a vigorous area of research [20–22].

There are some analytical methods to determine the presence of glyp and the metabolite AMPA in different media, such as water, urine, and serums. The official method to determine glyp in water, the EPA-547 [23], requires herbicide derivatization post-column with o-phathdialdehyde (OPA) [24]. Other methods employ high-performance liquid chromatography (HPLC) [25] coupled with mass spectrometry [26–29], fluorescence [30], or capillary electrophoresis [31]. However, these methods are complex, as they require pre-treatment steps for the samples and lengthy analysis times, by which it is not always possible to analyze massive samples in situ [32,33]. Recently, other methods have been reported as alternative tools for monitoring environmental samples in situ with the added benefit of short analysis times. Furthermore, some other analytical methods can have remarkably low detection limits, such as spectrophotometric [34–37], electrochemical [38–40], and Enzyme-Linked ImmunoSorbent Assay (ELISA) techniques [17,41]. The development of new methodologies that are quick, sensitive, reproducible, and inexpensive represent a viable alternative to the current methods and instruments by allowing the analysis of a larger number of samples either on the field or at the lab [42–44].

Dipyrromethenes are chelators of bipyrrolic monoacids that can form stable complexes with metals due to their coordination chemistry and optic and fluorescent properties [45]. These ligands are structurally rigid, completely conjugated, and capable of functionalizing several positions of their structure (1, 5, and 9) (Figure 1a). Hence, the coordination with metallic ions of meso-substituted dipyrromethenes (position 5) has been applied to the design of sensors. For instance, there is a study based on fluorescent probes on a boron dipyrromethene functionalized with a group of phenylboronic acids (BODIPY-PBAs) that can detect several monosaccharides in a concentration range of 0.1–100 mM, with good reproducibility and photostability [46]. In another study, an electrochemical biosensor was developed using a dipyrromethene-Cu(II) to determine the oligomeric form of amyloid beta (Aβ16-23) with concentrations in the range of 0.001–1.00 μM, which induces the neuronal dysfunction associated with Alzheimer´s disease (AD) [47]. Another paper reported on the electroactive dipyrromethene-Cu biosensor to detect antibodies against avian influenza virus type H5N1 in hen sera [48].

In this study, the chelating capacity of glyp was exploited to bind the metallic moiety of the compound bis 5-phenyldipyrrinate of nickel (II) ($Ni(PhDP)_2$) (Figure 1b). Based on this molecular association, a method for glyp determination in several water samples is developed for the quantification of the complex formed between glyp and ($Ni(PhDP)_2$). The resulting method is fast, sensitive, accurate, and useful for the quantification of glyp in drinking, urban, and ground waters, and residual-treated wastewater.

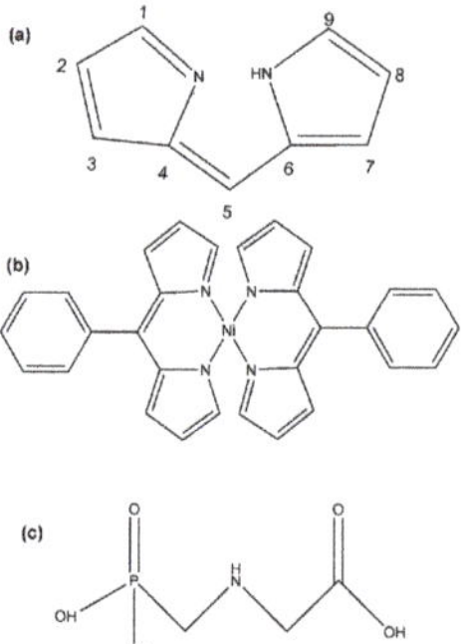

Figure 1. Chemical structure of (**a**) dipyrromethene ligand, (**b**) ($Ni(PhDP)_2$), and (**c**) glyp.

2. Materials and Methods

2.1. Reagents and Chemicals

Glyphosate, silica gel, pyrrole, trifluoroacetic acid (TFA), benzaldehyde, triethylamine (Et_3N), and 2,3-Dichloro-5,6-dicyano-1,4-benzoquinone (DDQ) were from Sigma-Aldrich (St. Louis, MO, USA). The methylene chloride (CH_2Cl_2), chloroform ($CHCl_3$), hexane, and methanol (CH_3OH) were from Fermont (Monterrey, Mexico), and ethyl acetate was purchased from JT Baker (Center Valley, PA, USA). All the used reagents are analytical grade.

2.2. Apparatus

Electronic absorbance spectra measurements were obtained using a Varian Cary 50 spectrophotometer equipped with a xenon lamp (Australia). To perform electronic spectra measurements, quartz cuvettes (1 cm path length) were used.

2.3. Synthesis of the ($Ni(PhDP)_2$)

The procedure used by Brückner et al. [49] was followed to synthesize ($Ni(PhDP)_2$). First, the 5-phenyldipyrrometane with benzaldehyde (1 mmol), pyrrole (1 mmol), and trifluoroacetic acid were synthesized. Then 2,3-Dichloro-5,6-dicyano-1,4-benzoquinone (1 mmol) was added and stirred for 30 min at room temperature. After that Et_3N (0.5 mL) was added to the solution and stirred for another 30 min at room temperature. After the elimination of the solvent, the residue was dissolved in CH_2Cl_2, and the solution was filtered to remove precipitates. The solvent was removed, and the product was purified by short silica gel column chromatography using ethyl acetate as an eluent. The first eluted yellow fraction was evaporated to afford the crude product. Later, 2 mmol of 5-Phenyldipyrromethene and 1 mmol of nickel sulfate hexahydrate were dissolved in a mixture of $CHCl_3$ and CH_3OH. The solution was agitated and heated under reflux for 6 h, and then 0.5 mL of Et_3N were added and heated again under reflux for 4 h. The solution was evaporated in a rotary evaporator until a dark brown solid was obtained. The dried remainder was dissolved in CH_2Cl_2 and CH_3OH (1:1), leaving it to slowly evaporate until crystals were obtained.

2.4. The Interaction between the ($Ni(PhDP)_2$) Compound and Glyp

The formation of the complex between ($Ni(PhDP)_2$) and glyp ($NiGlyp(PhDP)_2$) was followed by the changes in the electron absorption spectrum of ($Ni(PhDP)_2$) (1.08×10^{-4} mol/L) after the addition of glyp, in 1 mL of a 99% methanol, 1% water solution. A calibration curve was developed, registering the absorbance changes of the new band at 362 nm at different glyp concentrations (5.9×10^{-7} to 1.1×10^{-5} mol/L).

2.5. Complex Stoichiometry

The Job method of continuous variation [50] was used to determine the stoichiometry of ($NiGlyp(PhDP)_2$). Two equimolar stock solutions were prepared and mixed in a way that the total concentration was kept constant (5×10^{-5} and 1×10^{-4} mol/L for two different assays). The absorbance at 362 nm was measured after mixing for 10 s in a 1 mL reaction mixture (99% CH_3OH–1% water).

2.6. Determination of the Dissociation Constant (Kd)

To determine the dissociation constant, a curve was constructed, in which the absorbance changes at 362 nm of ($Ni(PhDP)_2$) (3.6×10^{-5} and 1.08×10^{-4} mol/L for two different assays) were monitored at different concentrations of glyp (5.9×10^{-7} to 2.3×10^{-4} mol/L), until its saturation was reached. The binding between the glyp and the ($Ni(PhDP)_2$) can be represented as Glyp + ($Ni(PhDP)_2$) $\rightarrow$ ($NiGlyp(PhDP)_2$).

The absorbance changes at 362 nm of the $(Ni(PhDP)_2)$ complex at different concentrations of glyp were transformed to a percentage of change and adjusted to the one-site binding model [51] (Equation (1)) to determine K_d:

$$\Delta AG = \Delta A_{max} \times K_d / G_0 \quad (1)$$

where ΔAG is the percentage of absorbance change at 362 nm upon adding each glyp concentration, ΔA_{max} is the maximum percentage of change (100%) when the $(Ni(PhDP)_2)$ is saturated with glyp, and G_0 is the total concentration of glyp. The reported values are the mean of the three replicates. The data were fit to the Hill equation using an iteration procedure following the Marquardt–Levenberg nonlinear least-squares algorithm, using Origin 8.0 software (Originlab Corporation, Northampton, MA, USA).

2.7. The Analysis in Water Samples

Four different water samples were analyzed to assess the potential of the methodology: potable, treated wastewater, urban, and groundwater, with each of them containing concentrations of glyp (4.1×10^{-6} and 5.9×10^{-6} mol/L) added intentionally. The analyses were carried out, in 1 mL of a 99% methanol, 1% water solution. Treated wastewater was filtered to remove suspended solids. The four water samples were stored at 4 °C until used and physicochemically characterized by following conventional methods: pH, specific conductance, Chemical Oxygen Demand (COD), and Biochemical Oxygen Demand (BOD). Several anion and cation analyses were performed as well: Ca^{2+}, Fe^{2+}, SO_4^{2-}, Mg^{2+}, NO^{3-}, NO_3-N, PO_4^{3-}, P_2O_5, and free chlorine.

2.8. Interfering Factors

To identify possible interfering factors, various salts were added at the maximum concentrations found in the environmental water samples: $FeCl_3 \cdot 6\ H_2O$ (6.6×10^{-4} mol/L showed 34 µs/cm), $CaCl_2 \cdot 2\ H_2O$ (9.8×10^{-4} mol/L showed 240 µs/cm), $NaNO_3$ (1.2×10^{-2} mol/L showed 140 µs/cm), and $MgCl_2 \cdot 6\ H_2O$ (4.9×10^{-5} mol/L showed 8 µs/cm), as well as the phosphate salts $Na_2PO_4 \cdot H_2O$ (140 µs/cm), $Na_2HPO_4 \cdot 7\ H_2O$ (140 µs/cm), and $(NH_4)_2HPO_4$ (146 µs/cm) at a concentration of 4.1×10^{-6} mol/L. Furthermore, other organophosphorus pesticides were tested: parathion (0.85 µs/cm), dimethoate (0.82 µs/cm), and diclophention (0.80 µs/cm), at a concentration of 4.1×10^{-6} mol/L for each one. The effect of the mix of salts (420 µs/cm) and pesticides (0.88 µs/cm) was also evaluated. For these studies, the assay time was 10 s, the glyp concentration of 4.1×10^{-6} mol/L (0.81 µs/cm) and the $(Ni(PhDP)_2)$ concentration of 1.08×10^{-4} mol/L were set.

All the experimental assays were made in triplicate to assess the repeatability of the results. The statistical analysis of the data was performed using Origin Software V 8.0.

2.9. Theoretical Structure of the $(NiGlyp(PhDP)_2)$ Complex

All calculations were performed using Density Functional Theory (DFT) [52,53] implemented in Gaussian 16 [54]. Geometry optimizations and frequency analysis were done in a vacuum with B3LYP functional and LANLD2Z basis sets. The local minima were identified with zero number of imaginary frequencies (NIMAG = 0). All calculations were made with no symmetry constraints.

3. Results and Discussion

3.1. Synthesis and Characterization of the $(Ni(PhDP)_2)$ Compound

The $(Ni(PhDP)_2)$ was successfully synthesized according to the procedure reported by Brückner et al. [49], and then it was characterized by UV-Vis (λ nm): 330, 480, and by mass spectrometry ^1H NMR spectroscopy based on the structural symmetry by signals for the aromatic protons. Xue et al. [55] characterized the $Ni(PhDP_2)$ with ^1H NMR to have (500 MHz, $CDCL_3$) $\delta = 9.426$ (s, 4, α-dipyrrin), 7.500

(d, J = 2.5 Hz, 6H, Ar-H), 7483–7.331 (m, 8H, Ar-H, β-dipyrrin), 6.741 (d, J = 3.5 Hz, 4H, β-dipyrrin), which corresponds with what was reported.

3.2. *The Interaction between Glyp with ($Ni(PhDP)_2$)*

As mentioned, the electronic absorption spectrum of ($Ni(PhDP)_2$) showed a characteristic band at 480 nm, indicative of double-bond electron transition metals (Figure 2, line), which significantly increased in the presence of glyp (1.1×10^{-5} mol/L); in addition, a new band at 362 nm was observed. The change in absorbance at 480 nm, as well as the formation of a new absorption band, is attributed to the formation of a complex between both species, with measurable characteristics in the ultraviolet-visible boundary region. The complex formation was almost instantaneous, and no additional changes in the absorbance were detected after mixing and measuring immediately (average time 10 seconds).

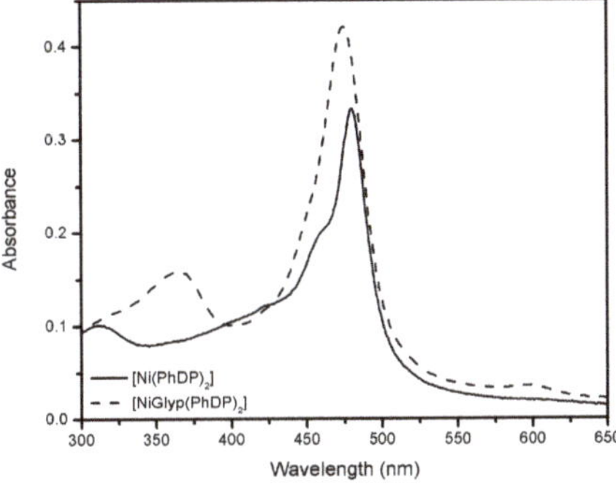

Figure 2. The electronic absorption spectrum of ($Ni(PhDP)_2$) (1.08×10^{-4} mol/L) in the absence (solid line) and the presence of glyp (1.1×10^{-5} mol/L) (dotted line). Reaction conditions: 99% CH_3OH–1% water.

Changes in the pH of the glyphosate solution, or the ($Ni(PhDPD)_2$) concentration, did not lead to improved results. Therefore, the pH of the water was kept at pH 7.0, ($Ni(PhDPD)_2$) of 1.08×10^{-4} mol/L, and incubation time of 10 s.

As the purpose of the present study is the quantification of glyp in water samples, the dependence of the absorbance change on glyp concentration was determined. As can be seen in Figure 3, the absorbance at 362 nm was dependent on the glyp concentration, with a linear range from 5.9×10^{-7} to 1.1×10^{-5} mol/L of glyp.

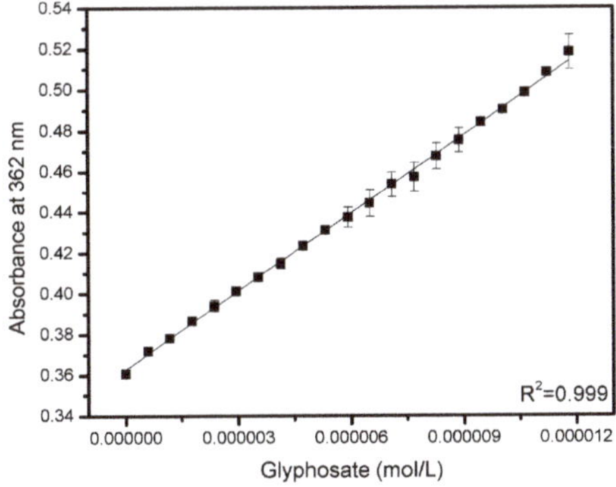

Figure 3. Calibration curve for the detection of glyp (5.9×10^{-7}–1.1×10^{-5} mol/L) based on the interaction with the compound ($Ni(PhDP)_2$) (1.08×10^{-4} mol/L). Reaction conditions: 99% CH_3OH–1% water.

Equation (2) describes the relationship between the two variables, where A is the absorbance at 362 nm, and G is the concentration of glyp (mol/L). The linear regression model fits the data very well, with a correlation coefficient higher than 0.99. Also, other statistical results (analysis of variance and dynamic graphs, i.e., normal probability plot of the standardized residuals, scatter plots of the standardized residual against the predictor variable, and the index plot of the standardized residuals) support the quality of the linear model.

$$A = 0.0128G + 0.3629 \quad (2)$$

Using Equation (2), the detection limit (LOD) and the quantification limit (LOQ) were calculated. The LOD and the LOQ are numerically equal to 3 and 10 times the standard deviation of the mean absorbance of ($Ni(PhDP)_2$) without glyp (blank absorbance). Substituting these values in Equation (1), a LOD of 2.01×10^{-7} mol/L (34.98 μg/L) and an LOQ of 9.87×10^{-7} mol/L (166 μg/L) were determined. These values are good enough for glyp determination in drinking water in the USA, as the EPA Maximum Contaminant Level (the highest level of a contaminant that is allowed in drinking water) of 700 μg/L has been set [56], and, in principle, should be higher for environmental water samples. A study by Botta et al. [57] reported concentrations of glyp of 4.4×10^{-7}–5.32×10^{-7} mol/L (75–90 μg/L) in surface water. Therefore, it is possible to find high concentrations of glyp in environmental water that are within the LOD and LOQ obtained in this work. However, an improvement in detection is needed for application of the method in European countries, where the maximum level of glyp must not be higher than 0.1 μg/L [58]. In addition, our LOD and LOQ values are within the range of other alternative methods to detect the herbicide in a water sample (Table 1), with the advantages that the response time is shorter and there is no need for derivatization or the addition of reaction precursors.

Table 1. Comparison between other glyp determination in water.

Method	LOD (mol/L)	LOQ (mol/L)	Remarks
This method	2.01×10^{-7}	9.87×10^{-7}	Rapid, effective, selective, facile, and sensitive
Spectrophotometry with multi-pumping flow system [59]	1×10^{-6}	3×10^{-6}	Rapid, effective and selective, but needs pre-treatments
Fluorescence resonance energy transfer [60]	6×10^{-7}	[1]	Rapid, effective and selective, but needs expensive equipment
Electrochemical sensing [43]	2×10^{-6}	[1]	Rapid, effective and selective, but needs expensive equipment and reagents
Colorimetric sensor [35]	6×10^{-7}	[1]	Effective and sensitive, but it requires complex synthetizing steps

Note: [1] Not reported.

3.3. Stoichiometry

For the complex stoichiometry, determination of the absorbance changes was plotted as a function of the mole fraction of the Glyp or ($Ni(PhDP)_2$) (Figure 4) at two total concentrations. The Job plots show a triangular shape, which according to literature suggests a strong molecular interaction between the compounds; also, the maximum point in the curves takes place at 0.5 mole fraction, indicating that the molecular association occurs with stoichiometry 1:1 [50,61].

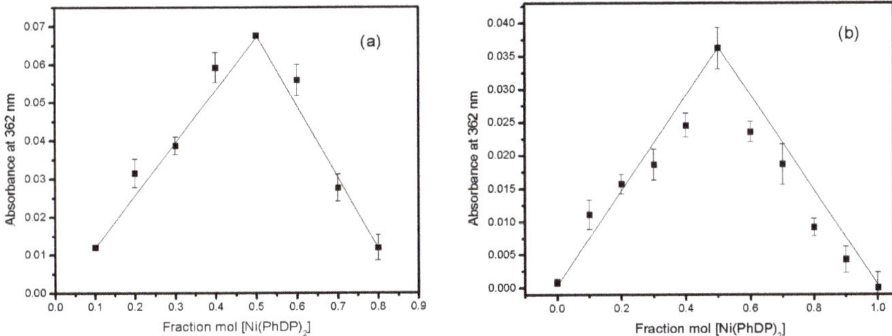

Figure 4. Job plot indicating the 1:1 stoichiometry for the (NiGlyp(PhDP)$_2$) complex, with a total concentration of (**a**) 5×10^{-5} mol/L and (**b**)1×10^{-4} mol/L.

3.4. Dissociation Constant

The K_d was determined by fitting the data to the one-site binding model using nonlinear regression analysis (Figure 5). The obtained K_d values were 1.75×10^{-6} and 6.9×10^{-6} mol/L at two (Ni(PhDP)$_2$) concentrations (3.6×10^{-5} and 1.08×10^{-4} mol/L), which account for the affinity of the dipyrrinate ligand to glyphosate. According to Chenprakhon et al. and Pan et al. [51,62], a small K_d value refers to a high binding affinity of the ligand to its target. Although the affinity values are three orders of magnitude lower compared to the antibody-antigen system, the dipirrinate ligand has the advantage that the analytic response does not need additional steps for separation and quantification, as in the immunoassay format.

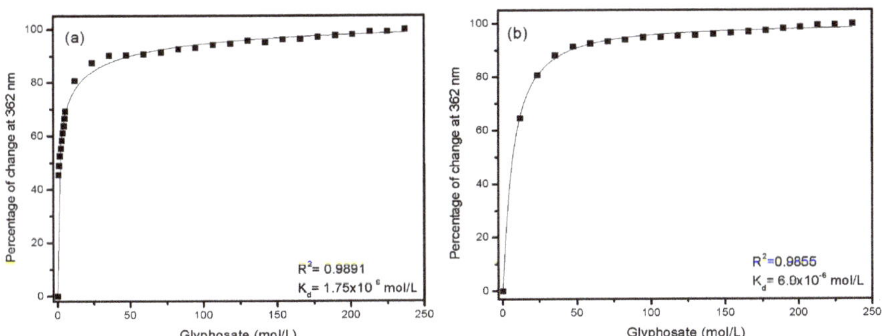

Figure 5. One site model fitting at (**a**) 1.08×10^{-4} mol/L of (Ni(PhDP)$_2$) and (**b**) 3.6×10^{-5} mol/L of (Ni(PhDP)$_2$) in the presence of glyp (5.9×10^{-7} to 2.36×10^{-4} mol/L).

3.5. Method Selectivity

The interference caused by other components that are usually present in water was determined to know the potential applicability of the method. It is well known that glyp can form strong coordination bonds with metal ions Fe^{2+}, Ca^{2+}, and Mg^{+2} [63,64]. The results showed that the presence of individual cations and its mixture did not interfere significantly in the complexation of glyp with (Ni(PhDP)$_2$). The same set of experiments were carried out with other organophosphorus pesticides, which are continuously used in agriculture, such as parathion [65], dimethoate [66], and diclofenthion [67], as well as their mixtures. The absorbance values of the (NiGlyp(PhDP)$_2$) complex in the presence of the organophosphorus pesticides and its mixtures did not show an interference greater than 10% and thus are discarded as interferents. A mixture was tested between (NiGlyp(PhDP)$_2$) and AMPA because AMPA is frequently detected in water together with glyp [68,69]. Interestingly, the metabolite showed no additional change in the detection of glyp. Finally, different phosphate salts were tested, as it is already known that the phosphate ion may form complexes with nickel and other metals [70–72], which also showed no interference. All the assays suggest a good selectivity of the method (Table 2).

Table 2. Effects of the added salts, organophosphorus pesticides, and phosphates on the method selectivity.

Compounds	Absorbance at 362 nm	Standard Deviation	Interference (%) [1]
None ($NiGlyp(PhDP)_2$)	0.42	0.012	0.0
	Salts		
$FeCl_3{\cdot}6H_2O$	0.43	0.008	2.57
$CaCl_2{\cdot}2H_2O$	0.41	0.008	0.19
$NaNO_3$	0.42	0.007	0.78
$MgCl_2{\cdot}6H_2O$	0.44	0.008	3.66
Mixture of salts	0.45	0.010	5.49
	Organophosphorus pesticides		
Parathion	0.42	0.012	1.47
Dimethoate	0.45	0.011	4.28
Diclofenthion	0.43	0.002	0.22
Mixture of pesticides	0.46	0.002	7.51
	Phosphates		
$Na_2PO_4{\cdot}H_2O$	0.41	0.003	2.57
$Na_2HPO_4{\cdot}7H_2O$	0.41	0.003	2.75
$(NH_4)_2HPO_4$	0.42	0.008	0.81
Mixture of phosphates	0.43	0.005	0.67
	Metabolite of glyp		
AMPA	0.42	0.12	0.0

[1] The percentage of interference was calculated, taking the absorbance of 0.42 obtained from the complex ($Ni(PhDP)_2$) 1.08×10^{-4} mol/L and glyp 4.1×10^{-6} mol/L as 0% of interference, minus the absorbance of the complex in the presence of interference between the absorbance of the complex, multiplied by one hundred.

3.6. Analysis of Spiked Water Samples

Table 3 shows the results of the physicochemical parameters of the four water matrices used to check the applicability of the proposed method. The parameter values are indicative of the different sources; as expected, treated wastewater shows the highest COD and BOD. The amount of dication metals was determined because of their possible interference behavior. No glyp was detected using the commercially available ELISA method (PN 500086) (Abraxis LLC, Warminster, PA, USA).

Table 3. Physicochemical analysis of the different water matrices.

Parameters	Water Matrix			
	Potable	Urban	Groundwater	Treated Wastewater
pH	7.00	7.00	7.00	7.00
Specific conductance (µs/cm)	0.05	424.00	523.00	1448.00
Temperature (°C)	25.00	25.30	25.30	24.70
COD (mg/L)	2.00	152.70	97.20	651.38
BOD (mg/L)	0.73	76.48	42.00	320.62
Ca^{2+} (mg/L)	20.00	75.00	115.00	145.00
Fe^{2+} (µg/L)	51.00	68.00	67.50	178.50
SO_4^{2-} (mg/L)	0.00	27.50	40.00	90.00
Mg^{2+} (mg/L)	0.00	10.00	30.0	10.00
NO_3-N (mg/L)	1.10	1.65	18.35	24.00
NO_3^- (mg/L)	5.00	82.75	99.45	104.00
PO_4^{3-} (mg/L)	0.10	4.80	0.94	40.20
P_2O_5 (mg/L)	0.08	4.53	0.67	31.00
Free chlorine (mg/L)	0.07	0.35	0.007	0.10

Then, the samples were spiked with glyp (4.1×10^{-6} and 5.9×10^{-6} mol/L) and added to the ($Ni(PhDP)_2$) solution. The glyp concentrations in the samples were calculated using Equation (2).

The percentages of recovery vary between 87.20–119.04% (Table 4), which indicates good accuracy of the methodology. Furthermore, a high precision was obtained, as reflected by the coefficients of variation (0.34–2.89%). Overall, it seems that the presence of several salts and metals at different concentrations did not affect the detection, and the method may be applied for different water sources. It is important to note that the LOD and LOQ reported here suggests that the method may be used for screening purposes in heavily polluted water samples, such as those in agricultural lands or treatment facilities, where the pollutants are concentrated. Regarding urban or groundwater, where pollutants are in lower concentrations, a preconcentration step should be necessary, as it is usually carried out with other methods.

Table 4. Detection of glyp in spiked water samples.

Water Matrix	Glyp Added ($\times 10^{-6}$ mol/L)	Glyp Determined ($\times 10^{-6}$ mol/L)	Coefficient of Variation (%)	Recovery (%)
Potable	4.10	3.72	2.89	89.58
	5.90	5.14	0.34	87.20
Urban	4.10	3.66	2.09	88.64
	5.90	5.32	0.61	89.97
Groundwater	4.10	4.02	2.38	96.49
	5.90	6.32	0.82	106.99
Treated wastewater	4.10	4.90	1.31	118.67
	5.90	7.03	0.99	119.04

3.7. Theoretical Results

To predict the possible complex between ($Ni(PhDP)_2$) and glyp quantum (QM), calculations were undertaken, starting the geometry optimization of probable compounds at the DFT level.

The structure complex shown in Figure 6 is the most probable compound found for the reaction of ($Ni(PhDP)_2$) with glyp, where the nickel in the ($NiGlyp(PhDP)_2$) complex expands its coordination number from 4 to 6 with a nitrogen and an oxygen atom of the glyp' carboxyl group, at distances between 1.89–2.08 Å in a distorted octahedral molecular geometry. The calculated bond distances are reported in Table 5. As can be seen, the distance values are similar to those reported in experimental conditions in the literature [55,70].

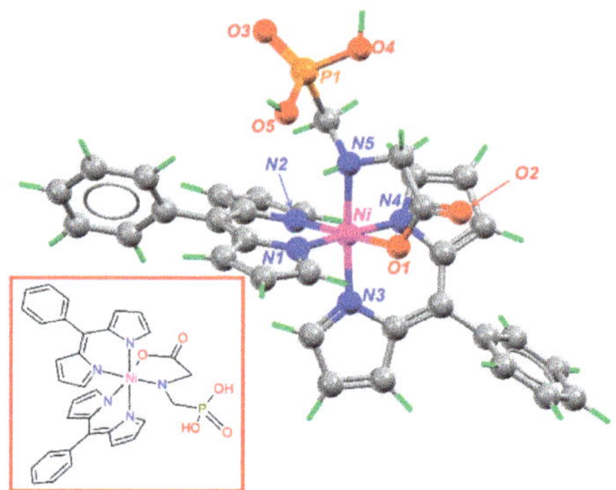

Figure 6. Optimized structure of the $(NiGlyp(PhDP)_2)$ complex.

The oxygen atoms of phosphonic did not interact with the nickel atom because during the optimization this complex is not stable. Although it is known that the glyp could coordinate to nickel in a tridentated fashion, it appears that a heptacoordinate compound, in this case, is not stable. The shorter distance of the oxygen of the carboxyl group indicates a stronger interaction of the glyp with the title compound.

Table 5. Bond distances (Å) for the coordination atoms of the optimized structure $(NiGlyp(PhDP)_2)$. Experimental values of $(Ni(PhDP)_2)$ fragment (**a**) and $Ni(Glyp)_2$ (**b**) from the literature are provided for comparison [55,70].

$(NiGlyp(PhDP)_2)$		$(Ni(PhDP)_2)$ (a) and $Ni(Glyp)_2$ (b)	
Bond	**Distance (Å)**	**Bond**	**Distance (Å)**
Ni-N(1)	1.93	Ni-N (a)	1.88
Ni-N(2)	1.93	Ni-N (b)	2.01
Ni-N(3)	1.94	Ni-O (b)	2.05
Ni-N(4)	1.90		
Ni-N(5)	2.08		
Ni-O	1.89		

4. Conclusions

It has been possible to detect glyp by complex formation with stoichiometry 1:1, achieving a LOD of 2.07×10^{-7} mol/L and an LOQ of 9.8×10^{-7} mol/L. The method developed assures data repeatability, high sensitivity, and quick detection (10 s).

The method was applied to determine known concentrations of glyp (4.1×10^{-6} and 5.9×10^{-6} mol/L) in potable and urban water, as well as groundwater and treated wastewater. The recovery percentages and the coefficients of variation obtained show good precision and accuracy of the method to be applied in environmental samples. The presence of the salts, other organophosphorus pesticides, and phosphates, as well as their mixtures in the water, do not interfere with glyp detection. The theoretical results show that the nickel of $(Ni(PhDP)_2)$ forms coordination bonds with the nitrogen and oxygen atoms of glyp in a distorted octahedral molecular geometry.

Author Contributions: A.R.-N. carried out the experiment, analysis and wrote the first draft of the paper; E.T. contributed to conducting a research and investigation process, oversight and leadership responsibility; I.P. contributed to verification of experiments and revision of the first draft; J.L.G.-M. contributed in synthesis process and preparation of published work; M.A. contributed to application of computational study and creation of molecular structure; E.G.-V. analyzed the study data and editing of publisher work.

Funding: This research project was funded by the Ph.D. Grant Number 494272 from Consejo Nacional de Ciencia y Tecnología (CONACYT).

Acknowledgments: Thanks to Marcela Ayala for allowing the use of UNAM supercomputing resources to perform the theoretical calculations, geometry optimizations, and frequency analysis (Project LANCAD-UNAM-DGTIC-293).

Conflicts of Interest: The authors declare no conflict of interest.

References

1. Dill, G.M.; Douglas, S.; Feng, P.C.C.; Kohn, F.; Kretzmer, K.; Mehrsheikh, A.; Bleeke, M.; Honegger, J.L.; Farmer, D.; Wright, D.; et al. Glyphosate: Discovery, development, applications, and properties. In *Glyphosate Resistance in Crops and Weeds: History, Development, and Management*; John Wiley & Sons: Hoboken, NJ, USA, 2010; pp. 1–34. ISBN 9780470410318.
2. Grube, A.; Donaldson, D.; Kiely, T.; Wu, L. *Pesticides Industry Sales, and Usage: 2006 and 2007 Market Estimates*; USA Environment Protection Agency: Washington, DC, USA, 2011; pp. 1–41.
3. Benbrook, C.M. Trends in glyphosate herbicide use in the United States and globally. *Environ. Sci. Eur.* **2016**, *28*, 1–42. [CrossRef] [PubMed]
4. Steinmann, H.H.; Dickeduisberg, M.; Theuvsen, L. Uses, and benefits of glyphosate in German arable farming. *Crop Prot.* **2012**, *42*, 164–169. [CrossRef]
5. Battaglin, W.A.; Meyer, M.T.; Kuivila, K.M.; Dietze, J.E. Glyphosate and its degradation product AMPA occur frequently and widely in U.S. soils, surface water, groundwater, and precipitation. *J. Am. Water Resour. Assoc.* **2014**, *50*, 275–290. [CrossRef]
6. Rendón-Von Osten, J.; Dzul-Caamal, R. Glyphosate residues in groundwater, drinking water and urine of subsistence farmers from intensive agriculture localities: A survey in Hopelchén, Campeche, Mexico. *Int. J. Environ. Res. Public Health* **2017**, *14*, 596. [CrossRef] [PubMed]
7. Wijekoon, N.; Yapa, N. Assessment of plant growth promoting rhizobacteria (PGPR) on potential biodegradation of glyphosate in contaminated soil and aquifers. *Groundw. Sustain. Dev.* **2018**, *7*, 456–469. [CrossRef]
8. Okada, E.; Costa, J.L.; Bedmar, F.R.; Ramsier, C.; Kloepper, J.W. Changes in rhizosphere bacterial gene expression following glyphosate treatment. *Sci. Total Environ.* **2016**, *553*, 32–41.
9. Krüger, M.; Schledorn, P.; Schrödl, W.; Hoppe, H.-W.; Lutz, W.; Shehata, A.A.D. Adsorption and mobility of glyphosate in different soils under no-till and conventional tillage. *Geoderma* **2016**, *263*, 78–85.
10. Degenhardt, D.; Humphries, D.; Cessna, A.J.; Messing, P.; Badiou, P.H.; Raina, R.; Farenhorst, A.; Pennock, D.J. Dissipation of glyphosate and aminomethylphosphonic acid in water and sediment of two Canadian prairie wetlands. *J. Environ. Sci. Health Part B Pestic. Food Contam. Agric. Wastes* **2012**, *47*, 631–639. [CrossRef]
11. Mercurio, P.; Flores, F.; Mueller, J.F.; Carter, S.; Negri, A.P. Glyphosate persistence in seawater. *Mar. Pollut. Bull.* **2014**, *85*, 385–390. [CrossRef] [PubMed]
12. Byer, J.D.; Struger, J.; Klawunn, P.; Todd, A.; Sverko, E.D. Low-cost monitoring of glyphosate in surface waters using the ELISA method: An evaluation. *Environ. Sci. Technol.* **2008**, *42*, 6052–6057. [CrossRef]
13. Çetin, E.; Şahan, S.; Ülgen, A.; Şahin, U. DLLME-spectrophotometric determination of glyphosate residue in legumes. *Food Chem.* **2017**, *230*, 567–571. [CrossRef] [PubMed]
14. Newman, M.M.; Lorenz, N.; Hoilett, N.; Lee, N.R.; Dick, R.P.; Liles, M. Detection of Glyphosate Residues in Animals and Humans. *J. Environ. Anal. Toxicol.* **2014**, *4*, 1–5.
15. Zouaoui, K.; Dulaurent, S.; Gaulier, J.M.; Moesch, C.; Lachatre, G. Determination of glyphosate and AMPA in blood and urine from humans: About 13 cases of acute intoxication. *Forensic Sci. Int.* **2013**, *226*, 20–25. [CrossRef] [PubMed]
16. Aparicio, V.C.; De Gerónimo, E.; Marino, D.; Primost, J.; Carriquiriborde, P.; Costa, J.L. Environmental fate of glyphosate and aminomethylphosphonic acid in surface waters and soil of agricultural basins. *Chemosphere* **2013**, *93*, 1866–1873. [CrossRef]
17. Mörtl, M.; Németh, G.; Juracsek, J.; Darvas, B.; Kamp, L.; Rubio, F.; Székács, A. Determination of glyphosate residues in Hungarian water samples by immunoassay. *Microchem. J.* **2013**, *107*, 143–151. [CrossRef]
18. Hu, Y.S.; Zhao, Y.Q.; Sorohan, B. Removal of glyphosate from aqueous environment by adsorption using water industrial residual. *Desalination* **2011**, *271*, 150–156. [CrossRef]

19. Ruiz-Toledo, J.; Castro, R.; Rivero-Pérez, N.; Bello-Mendoza, R.; Sánchez, D. Occurrence of glyphosate in water bodies derived from intensive agriculture in a tropical region of southern Mexico. *Bull. Environ. Contam. Toxicol.* **2014**, *93*, 289–293. [CrossRef] [PubMed]
20. Martínez, M.A.; Ares, I.; Rodríguez, J.L.; Martínez, M.; Martínez-Larrañaga, M.R.; Anadón, A. Neurotransmitter changes in rat brain regions following glyphosate exposure. *Environ. Res.* **2018**, *161*, 212–219. [CrossRef] [PubMed]
21. Valle, A.L.; Mello, F.C.C.; Alves-Balvedi, R.P.; Rodrigues, L.P.; Goulart, L.R. Glyphosate detection: Methods, needs and challenges. *Environ. Chem. Lett.* **2019**, *17*, 291–371. [CrossRef]
22. Van Bruggen, A.H.C.; He, M.M.; Shin, K.; Mai, V.; Jeong, K.C.; Finckh, M.R.; Morris, J.G. Environmental and health effects of the herbicide glyphosate. *Sci. Total Environ.* **2018**, *616–617*, 255–268. [CrossRef] [PubMed]
23. USA Environmental Protection Agency. *Method 547 Determination of Glyphosate in Drinking Water By Direct-Aqueous-Injection HPLC, Post-Column Derivatization, and Fluorescence Detection*; USA Environmental Protection Agency: Washington, DC, USA, 1990; Volume 1.
24. Kaczyński, P.; ozowicka, B. Liquid chromatographic determination of glyphosate and aminomethylphosphonic acid residues in rapeseed with MS/MS detection or derivatization/fluorescence detection. *Open Chem.* **2015**, *13*, 1011–1019. [CrossRef]
25. Kim, M.; Stripeikis, J.; Iñón, F.; Tudino, M. A simplified approach to the determination of N-nitroso glyphosate in technical glyphosate using HPLC with post-derivatization and colorimetric detection. *Talanta* **2007**, *72*, 1054–1058. [CrossRef]
26. Guo, Z.X.; Cai, Q.; Yang, Z. Determination of glyphosate and phosphate in water by ion chromatography-Inductively coupled plasma mass spectrometry detection. *J. Chromatogr. A.* **2005**, *1100*, 160–167. [CrossRef]
27. Ibáñez, M.; Pozo, Ó.J.; Sancho, J.V.; López, F.J.; Hernández, F. Re-evaluation of glyphosate determination in water by liquid chromatography coupled to electrospray tandem mass spectrometry. *J. Chromatogr. A* **2006**, *1134*, 51–55. [CrossRef] [PubMed]
28. Lee, E.A.; Strahan, A.P.; Thurman, E.M. *Methods of Analysis by the USA Geological Survey Organic Geochemistry Research Group—Determination of Glyphosate, Aminomethylphosphonic Acid, and Glufosinate in Water Using Online Solid-Phase Extraction and High-Performance Liquid Chromatography/Mass Sp*; Department of the Interior Washington: Washington, DC, USA, 2002.
29. Ibáñez, M.; Pozo, Ó.J.; Sancho, J.V.; López, F.J.; Hernández, F. Residue determination of glyphosate, glufosinate and aminomethylphosphonic acid in water and soil samples by liquid chromatography coupled to electrospray tandem mass spectrometry. *J. Chromatogr. A* **2005**, *1081*, 145–155. [CrossRef] [PubMed]
30. Padilla-Alonso, D.J.; Garza-Tapia, M.; Chávez-Montes, A.; González-Horta, A.; Waksman de Torres, N.H.; Castro-Ríos, R. New temperature-assisted ionic liquid-based dispersive liquid-liquid microextraction method for the determination of glyphosate and aminomethylphosphonic acid in water samples. *J. Liq. Chromatogr. Relat. Technol.* **2017**, *40*, 147–155. [CrossRef]
31. Moraes, M.P.; Gonçalves, L.M.; Pereira, E.A. Determination of glyphosate and aminomethylphosphonic acid by capillary electrophoresis with indirect detection using pyridine-2, 6-dicarboxylic acid or 3, 5-dinitrobenzoic acid. *Int. J. Environ. Anal. Chem.* **2018**, *98*, 258–270. [CrossRef]
32. Botero-Coy, A.M.; Ibáñez, M.; Sancho, J.V.; Hernández, F. Improvements in the analytical methodology for the residue determination of the herbicide glyphosate in soils by liquid chromatography coupled to mass spectrometry. *J. Chromatogr. A* **2013**, *1292*, 132–141. [CrossRef] [PubMed]
33. Yoshioka, N.; Asano, M.; Kuse, A.; Mitsuhashi, T.; Nagasaki, Y.; Ueno, Y. Rapid determination of glyphosate, glufosinate, bialaphos, and their major metabolites in serum by liquid chromatography-tandem mass spectrometry using hydrophilic interaction chromatography. *J. Chromatogr. A.* **2011**, *1218*, 3675–3680. [CrossRef]
34. Chang, Y.; Zhang, Z.; Hao, J.; Yang, W.; Tang, J. A simple label-free colorimetric method for glyphosate detection based on the inhibition of the peroxidase-like activity of Cu(II). *Sens. Actuators B Chem.* **2016**, *228*, 410–415. [CrossRef]
35. De Almeida, L.K.S.; Chigome, S.; Torto, N.; Frost, C.L.; Pletschke, B.I. A novel colorimetric sensor strip for the detection of glyphosate in water. *Sens. Actuators B Chem.* **2015**, *206*, 357–363. [CrossRef]

36. Rawat, K.A.; Majithiya, R.P.; Rohit, J.V.; Basu, H.; Singhal, R.K.; Kailasa, S.K. Mg^{2+} ion as a tuner for colorimetric sensing of glyphosate with improved sensitivity: Via the aggregation of 2-mercapto-5-nitrobenzimidazole capped silver nanoparticles. *RSC Adv.* **2016**, *6*, 47741–47752. [CrossRef]
37. Waiman, C.V.; Avena, M.J.; Garrido, M.; Fernández Band, B.; Zanini, G.P. A simple and rapid spectrophotometric method to quantify the herbicide glyphosate in aqueous media. Application to adsorption isotherms on soils and goethite. *Geoderma* **2012**, *170*, 154–158. [CrossRef]
38. Bettazzi, F.; Romero Natale, A.; Torres, E.; Palchetti, I. Glyphosate determination by coupling an immuno-magnetic assay with electrochemical sensors. *Sensors* **2018**, *18*, 2965. [CrossRef]
39. Sánchez-Bayo, F.; Hyne, R.V.; Desseille, K.L. An amperometric method for the detection of amitrole, glyphosate and its aminomethyl-phosphonic acid metabolite in environmental waters using passive samplers. *Anal. Chim. Acta* **2010**, *675*, 125–131. [CrossRef] [PubMed]
40. Songa, E.A.; Arotiba, O.A.; Owino, J.H.O.; Jahed, N.; Baker, P.G.L.; Iwuoha, E.I. Electrochemical detection of glyphosate herbicide using horseradish peroxidase immobilized on the sulfonated polymer matrix. *Bioelectrochemistry* **2009**, *75*, 117–123. [CrossRef] [PubMed]
41. Sanchís, J.; Kantiani, L.; Llorca, M.; Rubio, F.; Ginebreda, A.; Fraile, J.; Garrido, T.; Farré, M. Determination of glyphosate in groundwater samples using an ultrasensitive immunoassay and confirmation by on-line solid-phase extraction followed by liquid chromatography coupled to tandem mass spectrometry. *Anal. Bioanal. Chem.* **2012**, *402*, 2335–2345. [CrossRef]
42. Farkas, E.; Szekacs, A.; Kovacs, B.; Olah, M.; Horvath, R.; Szekacs, I. Label-free optical biosensor for real-time monitoring the cytotoxicity of xenobiotics: A proof of principle study on glyphosate. *J. Hazard. Mater.* **2018**, *351*, 80–89. [CrossRef] [PubMed]
43. Noori, J.S.; Dimaki, M.; Mortensen, J.; Svendsen, W.E. Detection of glyphosate in drinking water: A fast and direct detection method without sample pretreatment. *Sensors* **2018**, *18*, 2961. [CrossRef]
44. Vaghela, C.; Kulkarni, M.; Haram, S.; Aiyer, R.; Karve, M. A novel inhibition based biosensor using urease nanoconjugate entrapped biocomposite membrane for potentiometric glyphosate detection. *Int. J. Biol. Macromol.* **2018**, *108*, 32–40. [CrossRef]
45. Yamamura, M.; Albrecht, M.; Albrecht, M.; Nishimura, Y.; Arai, T.; Nabeshima, T. Red/near-infrared luminescence tuning of group-14 element complexes of dipyrrins based on a central atom. *Inorg. Chem.* **2014**, *53*, 1355–1360. [CrossRef] [PubMed]
46. Zhai, J.; Pan, T.; Zhu, J.; Xu, Y.; Chen, J.; Xie, Y.; Qin, Y. Boronic acid functionalized boron dipyrromethene fluorescent probes: Preparation, characterization, and saccharides sensing applications. *Anal. Chem.* **2012**, *84*, 10214–10220. [CrossRef]
47. Jargilo, A.; Grabowska, I.; Radecka, H.; Sulima, M.; Marszalek, I.; Wyslouch-Cieszyńska, A.; Dehaen, W.; Radecki, J. Redox-Active Dipyrromethene-Cu(II) Monolayer for Oriented Immobilization of His-Tagged RAGE Domains—the Base of Electrochemical Biosensor for Determination of Aβ16-23′. *Electroanalysis* **2013**, *25*, 1185–1193. [CrossRef]
48. Jarocka, U.; Sawicka, R.; Stachyra, A.; Góra-Sochacka, A.; Sirko, A.; Zagórski-Ostoja, W.; Sączyńska, V.; Porębska, A.; Dehaen, W.; Radecki, J.; et al. A biosensor based on electroactive dipyrromethene-Cu(II) layer deposited onto gold electrodes for the detection of antibodies against avian influenza virus type H5N1 in hen sera. *Anal. Bioanal. Chem.* **2015**, *407*, 7807–7814. [CrossRef] [PubMed]
49. Brückner, C.; Karunaratne, V.; Rettig, S.J.; Dolphin, D. Synthesis of meso-phenyl-4, 6-dipyrrins, preparation of their Cu(II), Ni(II), and Zn(II) chelates, and structural characterization of bis[meso-phenyl-4, 6-dipyrrinato]Ni(II). *Can. J. Chem.* **1996**, *74*, 2182–2193. [CrossRef]
50. Renny, J.S.; Tomasevich, L.L.; Tallmadge, E.H.; Collum, D.B. Method of continuous variations: Applications of job plots to the study of molecular associations in organometallic chemistry. *Angew. Chem. Int. Ed.* **2013**, *52*, 11998–12013. [CrossRef] [PubMed]
51. Chenprakhon, P.; Sucharitakul, J.; Panijpan, B.; Chaiyen, P. Measuring binding affinity of protein-ligand interaction using spectrophotometry: Binding of neutral red to riboflavin-binding protein. *J. Chem. Educ.* **2010**, *87*, 829–831. [CrossRef]
52. Hohenberg, P.; Kohn, W. The Inhomogeneous Electron Gas. *Phys. Rev.* **1964**, *136*, B864. [CrossRef]
53. Kohn, W.; Sham, L.J. Self-Consistent Equations Including Exchange and Correlation Effects *. *Phys. Rev.* **1965**, *140*, A1133–A1138. [CrossRef]

54. Frisch, M.J.; Trucks, G.W.; Schlegel, H.B.; Scuseria, G.E.; Robb, M.A.; Cheeseman, J.R.; Scalmani, G.; Barone, V.; Petersson, G.A.; Nakatsuji, H.; et al. *Gaussian 16*; Revision B.01; Gaussian. Inc.: Wallingford, CT, USA, 2016.
55. Wang, Y.; Xue, Z.; Dong, Y.; Zhu, W. Synthesis and electrochemistry of meso-substitutes dipyrromethene nickel (II) complexes. *Polyhedron* **2015**, *102*, 578–582. [CrossRef]
56. United States Environmental Protection Agency. *2018 Edition of the Drinking Water Standards and Health Advisories Tables*; United States Environmental Protection Agency: Washington, DC, USA, 2018.
57. Chevreuil, M.; Blanchoud, H.; Guery, B.; Moreau-Guigon, E.; Couturier, G.; Botta, F.; Alliot, F.; Fauchon, N.; Lavison, G. Transfer of glyphosate and its degradate AMPA to surface waters through urban sewerage systems. *Chemosphere* **2009**, *77*, 133–139.
58. Li, Z.; Jennings, A. Worldwide regulations of standard values of pesticides for human health risk control: A review. *Int. J. Environ. Res. Public Health* **2017**, *14*, 826. [CrossRef] [PubMed]
59. Silva, A.S.; Tóth, I.V.; Pezza, L.; Pezza, H.R.; Lima, J.L.F.C. Determination of glyphosate in water samples by multi-pumping flow system coupled to a liquid waveguide capillary cell. *Anal. Sci.* **2011**, *27*, 1031–1036. [CrossRef]
60. Yuan, Y.; Jiang, J.; Liu, S.; Yang, J.; Zhang, H.; Yan, J.; Hu, X. Fluorescent carbon dots for glyphosate determination based on fluorescence resonance energy transfer and logic gate operation. *Sens. Actuators B Chem.* **2017**, *242*, 545–553. [CrossRef]
61. Hill, Z.D.; MacCarthy, P. Novel approach to Job's method: An undergraduate experiment. *J. Chem. Educ.* **1986**, *63*, 162. [CrossRef]
62. Pan, Y.; Sackmann, E.K.; Wypisniak, K.; Hornsby, M.; Datwani, S.S.; Herr, A.E. Determination of equilibrium dissociation constants for recombinant antibodies by high-throughput affinity electrophoresis. *Sci. Rep.* **2016**, *6*, 1–11. [CrossRef]
63. Caetano, M.S.; Ramalho, T.C.; Botrel, D.F.; da Cunha, E.F.F.; de Mello, W.C. Understanding the inactivation process of organophosphorus herbicides: A DFT study of glyphosate metallic complexes with Zn^{2+}, Ca^{2+}, Mg^{2+}, Cu^{2+}, Co^{3+}, Fe^{3+}, Cr^{3+}, and Al^{3+}. *Int. J. Quantum Chem.* **2012**, *112*, 2752–2762. [CrossRef]
64. Ololade, I.A.; Oladoja, N.A.; Oloye, F.F.; Alomaja, F.; Akerele, D.D.; Iwaye, J.; Aikpokpodion, P. Sorption of Glyphosate on Soil Components: The Roles of Metal Oxides and Organic Materials. *Soil Sediment Contam.* **2014**, *23*, 571–585. [CrossRef]
65. Yola, M.L. Electrochemical activity enhancement of monodisperse boron nitride quantum dots on graphene oxide: Its application for simultaneous detection of organophosphate pesticides in real samples. *J. Mol. Liq.* **2019**, *277*, 50–57. [CrossRef]
66. Mondal, R.; Mukherjee, A.; Biswas, S.; Kole, R.K. GC-MS/MS determination and ecological risk assessment of pesticides in the aquatic system: A case study in Hooghly River basin in West Bengal, India. *Chemosphere* **2018**, *206*, 217–230. [CrossRef]
67. Ccanccapa, A.; Masiá, A.; Navarro-Ortega, A.; Picó, Y.; Barceló, D. Pesticides in the Ebro River basin: Occurrence and risk assessment. *Environ. Pollut.* **2016**, *211*, 414–424. [CrossRef] [PubMed]
68. Demonte, L.D.; Michlig, N.; Gaggiotti, M.; Adam, C.G.; Beldoménico, H.R.; Repetti, M.R. Determination of glyphosate, AMPA and glufosinate in dairy farm water from Argentina using a simplified UHPLC-MS/MS method. *Sci. Total Environ.* **2018**, *645*, 34–43. [CrossRef]
69. Fernandes, G.; Aparicio, V.C.; Bastos, M.C.; De Gerónimo, E.; Labanowski, J.; Prestes, O.D.; Zanella, R.; dos Santos, D.R. Indiscriminate use of glyphosate impregnates river epilithic biofilms in southern Brazil. *Sci. Total Environ.* **2019**, *651*, 1377–1387. [CrossRef] [PubMed]
70. Menelaou, M.; Dakanali, M.; Raptopoulou, C.P.; Drouza, C.; Lalioti, N.; Salifoglou, A. pH-Specific synthetic chemistry, and spectroscopic, structural, electrochemical and magnetic susceptibility studies in binary Ni(II)-(carboxy)phosphonate systems. *Polyhedron* **2009**, *28*, 3331–3339. [CrossRef]
71. Peleka, E.N.; Mavros, P.P.; Zamboulis, D.; Matis, K.A. Removal of phosphates from water by a hybrid flotation-membrane filtration cell. *Desalination* **2006**, *198*, 198–207. [CrossRef]
72. Subramaniam, V.; Hoggard, P.E. Metal Complexes of Glyphosate. *J. Agric. Food Chem.* **1988**, *36*, 1326–1329. [CrossRef]

Article

Oxazepam Alters the Behavior of Crayfish at Diluted Concentrations, Venlafaxine Does Not

Jan Kubec, Md Shakhawate Hossain, Kateřina Grabicová, Tomáš Randák, Antonín Kouba, Roman Grabic, Sara Roje and Miloš Buřič *

Faculty of Fisheries and Protection of Waters, South Bohemian Research Center of Aquaculture and Biodiversity of Hydrocenoses, University of South Bohemia in České Budejovice, Zátiší 728/II, 389 25 Vodňany, Czech Republic; kubecj@frov.jcu.cz (J.K.); mhossain@frov.jcu.cz (M.S.H.); grabicova@frov.jcu.cz (K.G.); trandak@frov.jcu.cz (T.R.); akouba@frov.jcu.cz (A.K.); rgrabic@frov.jcu.cz (R.G.); sroje@frov.jcu.cz (S.R.);

* Correspondence: buric@frov.jcu.cz; Tel.: +420-387-774-769

Received: 11 December 2018; Accepted: 20 January 2019; Published: 24 January 2019

Abstract: Pharmaceutically active compounds are only partially removed from wastewaters and hence may be major contaminants of freshwaters. Direct and indirect effects on aquatic organisms are reported at dilute concentrations. This study was focused on the possible effects of environmentally relevant concentrations (~1 μg L^{-1}) of two psychoactive compounds on the behavior of freshwater crayfish. Experimental animals exposed to venlafaxine did not show any behavioral alteration. Crayfish exposed to the benzodiazepine oxazepam exhibited a significant alteration in the distance moved and activity, and the effects were different when individuals were ready for reproduction. Results suggested that even the low concentration of selected psychoactive pharmaceuticals could alter the behavioral patterns of crayfish, as reported for other pharmaceuticals. These results provide new information about the possible adverse effects of pharmaceuticals at dilute concentrations. From previous knowledge and our results, it is obvious that different compounds have different effects and the effects are even specific for different taxa. Detailed studies are therefore needed to assess the possible ecological consequences of particular substances, as well as for their mixtures.

Keywords: environmental pollution; pharmaceuticals; freshwaters; crayfish

1. Introduction

Pharmaceutically active compounds (PhAC) are considered emerging contaminants in aquatic environments [1,2]. PhACs originate mainly from human or animal excretion or runoff from hospitals [3,4] and penetrate freshwaters via effluents of sewage treatment plants (STPs) which are ineffective in their removal [5]. The residues have several non-lethal effects on aquatic organisms and, through them, on whole ecosystems [6,7]. Psychotropic substances are present often at much lower concentrations in surface waters [8–10] than, for example, antibiotics or hypertension drugs, [11,12] but they have important effects in very diluted concentrations as well [13,14].

The psychotropic substances venlafaxine and oxazepam alter the state of the brain by flooding it with the neurotransmitter serotonin (5-HT) or act on benzodiazepine receptors, having direct inhibitory effects on the central nervous system [15,16]. Invertebrates (including crayfish) have similar receptors for psychotropic compounds as mammals [17], even with the potential for bioaccumulation [14,18], which increases the probability of the apparent PhACs' effects on these animals.

Some psychoactive PhACs are also bioactive and can persist in the sediments of surface waters [19], enabling their transfer via the food-web [20]. They are developed to modify behavioral patterns, so a behavioral alteration in aquatic organisms is likely [21,22]. However, the behavioral effects of these psychotropic compounds still remain less understood than eco-toxicity tests [23]. Behavioral effects, from an ecological point of view, can affect the survival of an individual, as well as the long-term sustainability of a population [24]. Crayfish seem to be good model organisms, having well known social and spatial behavior [25,26] and being similarly susceptible to the behavioral changes induced by PhACs [13,27,28].

In this study, the behavioral patterns of a clonal species, the marbled crayfish (*Procambarus virginalis*, Lyko 2017), were assessed using an ethological software where control animals and those exposed to environmentally relevant concentrations of venlafaxine and oxazepam were used. We hypothesized about the possible behavioral changes associated with the pollutants used at concentrations commonly detected in surface waters, as confirmed in our previous study with other PhACs.

2. Materials and Methods

2.1. Chemicals

Venlafaxine (VEN) and oxazepam (OXA) were obtained from AK Scientific (Union City, CA, USA) and Lipomed (Cambridge, MA, USA), respectively. Stock solutions of both compounds (concentration of 10 mg L^{-1}) were prepared using ultra-pure water (AquaMax Basic 360 Series and Ultra 370 Series instrument, Young Lin, Anyang, Republic of Korea) and were stored at 4 °C. The exposure solutions of 1 μg L^{-1} were prepared by dilution of the stock solution in aged tap water. Concentration testing was utilized to evaluate reported [29] environmentally relevant concentrations.

Isotopically labeled venlafaxine (D6-VEN) and oxazepam (D5-OXA; both from Lipomed (USA)) were used as internal standards for the analyses of water samples. Ultra-pure water and acetonitrile (LC/MS grade purity, Merck, Kenilworth, NJ, USA), both acidified with formic acid (Sigma-Aldrich, Darmstadt, Germany), were used as the mobile phases in liquid chromatography (LC).

2.2. Experimental Animals

Marbled crayfish (with a carapace length of 16–22 mm, measured using a vernier caliper to the nearest 0.1 mm) were randomly selected from our laboratory culture. Crayfish weight (to the nearest 0.1 g) was obtained using an electronic balance (Kern & Sohn GmbH, Balingen, Germany). The mean lengths and weights of the animals used (Table 1) did not differ between the control and exposed groups.

Table 1. The mean carapace length (CL) and weight (W) of marbled crayfish (*Procambarus virginalis*, Lyko 2017) animals used in the experimental groups in either the presence or absence of shelter. Data are presented as mean ± standard deviation. The *t*-test values and *p*-values are shown to demonstrate no differences between experimental groups.

Tested Compound	Group	Shelter	CL (mm)	*t*-test	*p*	W (g)	*t*-test	*p*
Venlafaxine	exposed	no	19.0 ± 1.8	−0.13	0.897	2.0 ± 0.6	−0.14	0.889
	control	no	18.8 ± 2.1			1.9 ± 0.7		
	exposed	yes	18.3 ± 2.1	0.39	0.695	1.8 ± 0.7	0.39	0.699
	control	yes	18.4 ± 2.0			1.8 ± 0.7		
Oxazepam	exposed	no	20.1 ± 3.0	1.12	0.269	2.3 ± 0.9	1.39	0.171
	control	no	19.3 ± 2.3			2.0 ± 0.8		
	exposed	yes	19.1 ± 2.8	0.36	0.717	2.1 ± 0.9	0.43	0.672
	control	yes	18.8 ± 2.7			2.0 ± 0.9		

2.3. Experimental Design

The exposition and experimental work was conducted in November (VEN) and December (OXA) 2017. In total, 55 crayfish were exposed to a concentration of ~1 μg L^{-1} of VEN for 21 days and 60 animals to an OXA compound for 7 days, respectively. The concentration was chosen based on previously reported environmental concentrations [9,14,18,30,31]. The exposure times were chosen in relation to the mode of action of the selected compounds. VEN acts when a steady-state plasma concentration is achieved (3–4 weeks) [32], while OXA acts immediately [16]. Crayfish maintained in aged tap water were used as controls, with the same handling as the exposed groups. The crayfish were held individually in transparent plastic boxes (190 × 140 × 75 mm) with 0.5 L of exposure solution or aged tap water. The numbers of animals that molted, spawned, or died during the exposure period were recorded.

During the exposure period, crayfish were fed ad libitum with fish pellets (Sera Granugreen, Sera, Heinsberg, Germany). Boxes were cleaned during exposure to the solution and the water exchange (every 48 h). The control group was always handled first to avoid its contamination. Crayfish which molted or spawned were discarded from the experiment. Water temperature was measured by an alcohol thermometer (to the nearest 0.1 °C) and did not differ ($p > 0.05$) between the control and the exposed group in both VEN and OXA studies. Water temperature ranged between 19.3 and 20.6 °C.

The real concentrations of VEN and OXA in the exposure solution, as well as in the control group's water, was checked using liquid chromatography with a tandem mass spectrometer (LC-MS/MS, Research Institute of Fish Culture and Hydrobiology, Vodňany, Czech Republic) four (VEN) and three times (OXA) during the exposure period. The concentrations of the compounds were analyzed in freshly prepared solutions (at time 0) and after 48 h, when the used solution was exchanged (time 48). Collected samples were filtered (0.20 μm regenerated cellulose, Labicom, Olomouc, Czech Republic) and stored in a freezer at −20 °C until analysis. After thawing and the addition of the internal standards, the samples were measured using the 10 min method on a Hypersil Gold aQ column (50 × 2.1 mm; 5 mm particles) coupled with an Accela 1250 LC pump and a TSQ Quantum Ultra Mass Spectrometer (Thermo Fisher Scientific, Waltham, MA, USA). The concentrations of the tested compounds in the analyzed water samples from the exposed boxes at time 0 and time 48 did not differ. The concentrations of VEN and OXA in water samples from the control group were below the limit of detection (see Table 2).

Table 2. The concentration of VEN and OXA in the water at time 0 (control, exposed) and after 48 h of exposure (control, exposed) ($\alpha = 0.05$). Data are presented as mean ± standard deviation.

Tested Compound	Group	n	Time 0 ($\mu g\ L^{-1}$)	Time 48 ($\mu g\ L^{-1}$)	Paired *t*-test	*p*
Venlafaxine	exposed	4	0.7 ± 0.0	0.7 ± 0.0	0.927	0.42
	control	4	<0.02	<0.02	---	---
Oxazepam	exposed	3	1.3 ± 0.3	1.2 ± 0.2	0.983	0.43
	control	3	<0.01	<0.01	---	---

2.4. Behavioral Data Acquisition

The exposed crayfish were individually placed in circular plastic tanks (280 mm in diameter), with 2 L of aged tap water and 200 mL of fine sand (<1 mm). In total, 110 and 120 crayfish were used for video-tracking in the VEN and OXA experiments, respectively. Stocked crayfish were video-tracked using a digital video camera (Sony HDR-CX240, Sony, Tokyo, Japan) in trials of 20 parallel tracked tanks (10 control and 10 exposed animals), i.e., 6 trials were done for each compound. Half of the trials were made without shelter, while the other half were conducted with shelter (consisting of halved ceramic plant pots of a 60 mm entry width and a depth of 50 mm). Shelter is an essential resource of crayfish, being nocturnal animals which are usually only active for a period throughout the day, and affecting shelter use can make crayfish more prone to predation or cannibalism. After the video recording, the presence of glair glands (a mark of readiness for reproduction) was also recorded in the used crayfish, due to the possible consequences of upcoming reproduction on their behavior.

The video-recording lasted for 4 h. Light was provided, as permanent indirect illumination, by fluorescent tubes (daylight, 2310 lm). Video-recordings were analyzed later using EthoVision® XT 13.0 software (Noldus Information Technology by Wageningen, The Netherlands) with a multiple-arena module. The distance moved (cm), activity (percentage of time when crayfish locomotion was detected), and velocity (cm s^{-1}) were evaluated. When conditions of shelter were present, the software also revealed results about the time spent inside/outside the shelter.

2.5. Statistical Analysis

A chi-square test was used for numbers of molted and spawned crayfish in comparison with control ones. The Kolmogorov-Smirnov normality test was done for the entire data set. The homogeneity of variances was tested using the Bartlett test for the parameters of behavioral patterns. The concentrations of the tested compound at time 0 and time 48 were compared through paired *t*-tests. A *t*-test for independent samples was used to compare the size and weight of the animals used in the exposed and control groups. The distance moved, velocity, activity, and time spent outside the shelter (replicate groups as a random factor, exposure as a fixed factor), were analyzed through factorial ANOVA. The null hypothesis was rejected at $\alpha = 0.05$. The data were statistically analyzed by Statistica 12.0 software (StatSoft, Tulsa, OK, USA).

3. Results

3.1. Venlafaxine

No significant differences were detected in VEN-exposed crayfish in comparison with control animals in set-ups both with and without available shelter (Table 3). The only effect detected was that of developed glair glands in the set-up with shelter on activity ($F_{1,44} = 6.95$, $p = 0.012$) and time spent outside the shelter ($F_{1,44} = 4.94$, $p = 0.031$). No values are recorded for crayfish without shelter due to the absence of glair glands (Table 4). Data are shown in Tables 3 and 4.

Table 3. The values of the distance moved, velocity, activity, and time spent outside the shelter in crayfish exposed to venlafaxine and in control crayfish, in set-ups with and without available shelter. Data are shown as mean ± standard deviation.

Compound	Shelter Available	Distance Moved (cm)	Velocity (cm s^{-1})	Activity (%)	Time Spent Outside the Shelter (%)
Venlafaxine	no	5184 ± 367	0.39 ± 0.03	69.9 ± 1.8	---
Control	no	5195 ± 382	0.38 ± 0.03	67.8 ± 2.2	---
Venlafaxine	yes	1569 ± 339	0.65 ± 0.07	59.3 ± 5.9	27.0 ± 5.5
Control	yes	1618 ± 363	0.67 ± 0.05	66.0 ± 5.5	24.9 ± 6.5

Table 4. The values of the distance moved, velocity, activity, and time spent outside the shelter in crayfish exposed to venlafaxine and in control crayfish in accordance with the presence of glair glands, in set-ups with and without available shelter. Data are shown as mean ± standard deviation.

<table>
<tr><th>Compound</th><th>Shelter Available</th><th>Glair Glands</th><th>Distance Moved (cm)</th><th>Velocity (cm s^{−1})</th><th>Activity (%)</th><th>Time Spent Outside the Shelter (%)</th></tr>
<tr><td rowspan="2">Venlafaxine</td><td rowspan="4">no</td><td>yes</td><td>---</td><td>---</td><td>---</td><td>---</td></tr>
<tr><td>no</td><td>5184 ± 367</td><td>0.39 ± 0.03</td><td>69.9 ± 1.8</td><td>---</td></tr>
<tr><td rowspan="2">Control</td><td>yes</td><td>---</td><td>---</td><td>---</td><td>---</td></tr>
<tr><td>no</td><td>5195 ± 382</td><td>0.38 ± 0.03</td><td>67.8 ± 2.2</td><td>---</td></tr>
<tr><td rowspan="2">Venlafaxine</td><td rowspan="4">yes</td><td>yes</td><td>1397 ± 571</td><td>0.78 ± 0.10</td><td>76.0 ± 5.3</td><td>14.8 ± 6.1</td></tr>
<tr><td>no</td><td>1676 ± 542</td><td>0.62 ± 0.17</td><td>48.9 ± 8.0</td><td>34.7 ± 7.6</td></tr>
<tr><td rowspan="2">Control</td><td>yes</td><td>1279 ± 693</td><td>0.92 ± 0.17</td><td>76.4 ± 7.8</td><td>12.3 ± 8.3</td></tr>
<tr><td>no</td><td>1777 ± 660</td><td>0.51 ± 0.07</td><td>61.1 ± 7.5</td><td>30.7 ± 8.5</td></tr>
</table>

3.2. Oxazepam

In the set-up without available shelter, crayfish moved longer distances ($F_{1,56} = 4.17$, $p = 0.046$) and showed higher activity ($F_{1,56} = 4.75$, $p = 0.034$) when exposed to OXA, compared to the control group. The effect of glair glands was observed only in the control animals in both the distance moved ($F_{1,56} = 5.73$, $p = 0.024$) and activity ($F_{1,56} = 6.92$, $p = 0.013$). In OXA-exposed animals, this effect was not so obvious and no differences in either the distance moved ($F_{1,56} = 0.20$, $p = 0.656$) or activity ($F_{1,56} = 1.36$, $p = 0.248$) were revealed. Similarly, no significant difference was revealed in the velocity, either between the control and exposed animals or between the animals with and without glair glands (Figures 1 and 2).

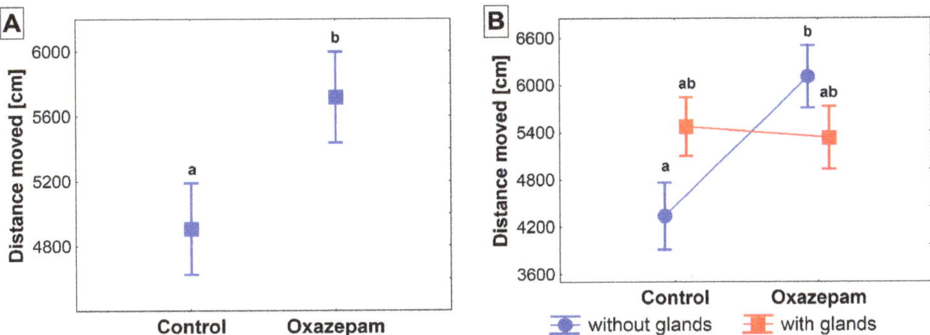

Figure 1. The distance moved (in cm) of marbled crayfish (*Procambarus virginalis*, Lyko 2017) exposed to oxazepam (~1 µg L^{-1}) and of the control animals in the conditions without available shelter (**A**). The differences detected between the groups of crayfish and in accordance with the presence of developed glair glands (**B**). The different superscripts show significant differences ($\alpha = 0.05$) among groups. Data are presented as mean ± standard error of mean.

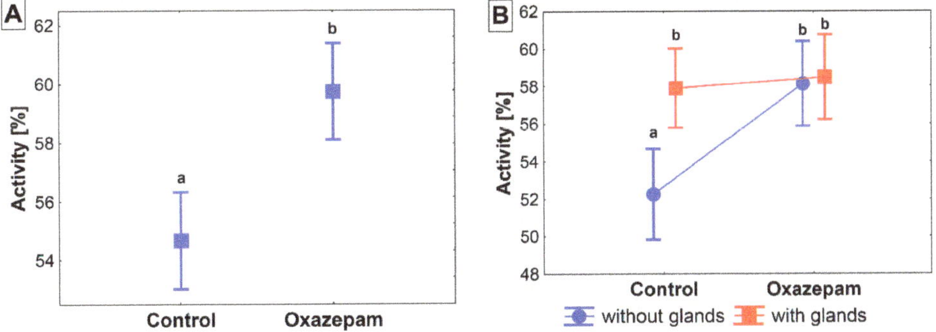

Figure 2. The activity (in %) of marbled crayfish (*Procambarus virginalis*, Lyko 2017) exposed to oxazepam (~1 µg L^{-1}) and control crayfish in the conditions without available shelter (**A**). The differences detected between the groups of crayfish in accordance with the presence of developed glair glands (**B**). The different superscripts show significant differences ($\alpha = 0.05$) among groups. Data are presented as mean ± standard error of mean.

In the set-up with available shelter, no differences were detected between the control and OXA-exposed animals. Only the effect of glair glands was detected in all of the parameters observed; the distance moved ($F_{1,56} = 8.74$, $p = 0.005$), velocity ($F_{1,56} = 4.24$, $p = 0.044$), activity ($F_{1,56} = 9.47$, $p = 0.003$), and the time spent outside the shelter ($F_{1,56} = 6.51$, $p = 0.014$).

Data are summarized in Tables 5 and 6.

Table 5. The values of the distance moved, velocity, activity, and time spent outside the shelter in crayfish exposed to oxazepam and in control crayfish, in set-ups with and without available shelter. Data are shown as mean ± standard deviation.

Compound	Shelter Available	Distance Moved (cm)	Velocity (cm s^{-1})	Activity (%)	Time Spent Outside the Shelter (%)
Oxazepam	yes	1848.1 ± 312.7	0.39 ± 0.03	18.7 ± 2.2	29.3 ± 3.2
Control	yes	1852.6 ± 256.8	0.43 ± 0.03	18.6 ± 2.6	28.9 ± 3.9

Table 6. The values of the distance moved, velocity, activity, and time spent outside the shelter in crayfish exposed to oxazepam and in control crayfish in accordance with the presence of glair glands, in set-ups with and without available shelter. Data are shown as mean ± standard deviation.

Compound	Shelter Available	Glair Glands	Distance Moved (cm)	Velocity (cm s^{-1})	Activity (%)	Time Spent Outside the Shelter (%)
Oxazepam	no	yes	5515.2 ± 360.2	0.39 ± 0.02	59.5 ± 2.9	---
		no	5967.7 ± 642.7	0.43 ± 0.05	60.2 ± 2.4	---
Control		yes	5471.6 ± 378.1	0.40 ± 0.03	52.3 ± 1.8	---
		no	4570.5 ± 160.8	0.35 ± 0.02	57.9 ± 2.1	---
Oxazepam	yes	yes	2382.9 ± 606.3	0.42 ± 0.04	22.8 ± 3.9	34.1 ± 5.4
		no	1380.2 ± 210.7	0.35 ± 0.03	15.1 ± 2.2	25.1 ± 3.6
Control		yes	2483.9 ± 405.1	0.47 ± 0.04	24.7 ± 4.4	36.8 ± 6.7
		no	1221.2 ± 228.4	0.38 ± 0.03	12.5 ± 2.0	21.0 ± 3.0

3.3. *Life History Traits*

During the exposure period, the number of molted crayfish did not vary significantly in both VEN ($\chi^2 = 0.08$, $p = 0.783$) and OXA ($\chi^2 = 0.72$, $p = 0.398$) exposed crayfish compared to control ones. However, the number of crayfish which spawned eggs during the exposure period was slightly lower but not statistically significant ($\chi^2 = 2.81$, $p = 0.094$) in the VEN-exposed group compared to the control group. The number of spawned animals in the OXA-exposed group did not differ from the controls ($\chi^2 = 0.15$, $p = 0.703$). There was no reported mortality (Table 7).

Table 7. The total number of molted, spawned, and dead crayfish in the control and exposed groups during the exposure period to venlafaxine and oxazepam.

Pharmaceutical	Group	Molted (n)	Spawned (n)	Dead (n)
Venlafaxine	control	61	4	0
	exposed	58	1	0
Oxazepam	control	12	3	0
	exposed	9	4	0

4. Discussion

Pharmaceuticals accessing natural ecosystems via sewage water treatment plant effluents [5,33] are reported as drivers of ecological changes [18,34], and psychotropic drugs are often confirmed as inducing behavioral changes in fish and other aquatic invertebrates [13,35]. Knowledge about the behavioral endpoint of these drugs is still too scarce to summarize their ecological consequences [13].

The present study assessed the behavioral effects of two psychoactive compounds, VEN and OXA, on clonal marbled crayfish exposed to dilute concentrations which can be found in natural conditions. The study also follows up on previous studies [13,36] exploring the compound-specific effects on the behavior of a model invertebrate in comparable, defined conditions. The results observed again confirmed that the low concentrations observed in natural conditions can have important consequences. Exploratory behavior, expressed as the activity and distance moved, should affect the wasting of energy, leading to a shorter life span in more active individuals [37]. In addition, the visibility of an animal to predators, especially in conditions of invaded ecosystems, can have effects on food resources, which are under greater pressure due to the higher activity [38,39]. Such alterations can then change the ecosystem's functioning [40,41]. The elevated activity, followed by altered foraging behavior can, therefore, lead to the breakdown of food chains, loss of biodiversity, and ecosystem instability [41,42].

Our experimental data helped to identify the environmental risks of OXA, but surprisingly no behavioral changes have been observed for the antidepressant VEN. Venlafaxine effects are reported at higher concentrations [43] compared to other antidepressants, citalopram [13] and sertraline [36], which were tested at the same (or even lower concentrations in the case of fluoxetine and sertraline) [44].

Earlier research demonstrated different, or even opposite results among similar compounds or compounds with similar modes of action, as well as among different taxa observed [21,35,45]. This can be due, in part, to different experimental conditions, ways of application (injection, oral application, passively from a water solution), doses (lower than in environmentally detected, environmentally relevant, elevated), and approaches used to determine the behavioral effects. We, therefore, tried to minimize the variables and use similar, relatively easy approaches for the observation, an application imitating the environmental intoxication from the dilute solutions, and the use of genetically uniform marbled crayfish, to erase the effect of different genotypes. However, the results presented, compared with previous ones produced with the same methodology, reveal again the differences among individual compounds. In fact, there is a need to investigate the mechanisms of action of these substances in detail and to elaborate on studies dealing with different mixtures of pharmaceuticals as they act simultaneously on aquatic biota [46].

The present study can be helpful for the set-up of new studies aimed not only at behavioral patterns but at life history traits, including reproduction. Behavioral alterations provide the potential to assess the risk of ecological effects of pharmaceutical products, thus it might be useful for the generalization of the impacts on the aquatic environment. In our present study, we also observed changed behavior in crayfish with developed glair glands, indicative of the preparation for reproduction [37,47]. To safeguard their future offspring, crayfish limit their activity, which was expressed as a lower distance moved and less activity. In the exposed animals, this pattern was missing, which can affect the reproductive success in a population due to a higher risk of predation. Life history traits and affected reproduction have also been confirmed from other studies with several psychoactive drugs on different aquatic taxa [36,43,48].

To summarize, our results together with those previously published show high variability in the type, strength, and direction of the effects of pharmaceuticals on aquatic biota. The use of passive application due to exposure through diluted, environmentally relevant concentrations of tested compounds seems to be appropriate for the assessment of their environmental effects. The main pollutants should also be tested individually, as well as their mixtures as found in field sampling [46]. The large number of compounds in surface waters and their different/specific modes of action is motivation for the better understanding of their real ecological impact on ecosystems.

Author Contributions: Conceptualization, T.R., R.G. and M.B.; formal analysis, J.K.; funding acquisition, T.R.; investigation, J.K., M.S.H., A.K., S.R. and M.B.; methodology, K.G., T.R., A.K. and M.B.; project administration, T.R.; resources, K.G., T.R. and R.G.; supervision, T.R., R.G. and M.B.; validation, K.G., R.G. and M.B.; visualization, J.K. and M.S.H.; writing—original draft, J.K., M.S.H., A.K. and M.B.; writing—review and editing, K.G., A.K. and M.B.

Funding: The study was financially supported by the Czech Science Foundation (project No. 16-06498S), the Ministry of Education, Youth and Sports of the Czech Republic—projects CENAKVA (No. CZ.1.05/2.1.00/01.0024), CENAKVA II (No. LO1205 under the NPU I program), CENAKVA Center Development (No. CZ.1.05/2.1.00/19.0380) and by the Grant Agency of University of South Bohemia No. 012/2016/Z.

Acknowledgments: We would like to thank our colleagues Wei Guo and Filip Ložek who helped us during experimental work. We also deeply appreciate the help of Julian D. Reynolds, not only in language corrections of the manuscript but for friendly mentoring too.

Conflicts of Interest: The authors declare no conflict of interest. The funders had no role in the design of the study; in the collection, analyses, or interpretation of data; in the writing of the manuscript, and in the decision to publish the results.

Ethical code: No specific permissions were required for the locations and activities involved in this study. All experiments were conducted according to the principles of the Ethics Committee for the Protection of Animals in Research of the University of South Bohemia in České Budějovice, FFPW, Vodňany, based on the EU-harmonized Animal Welfare Act of the Czech Republic, No. 53100/2013-MZE-17214.

References

1. Burkina, V.; Zlabek, V.; Zamaratskaia, G. Effects of pharmaceuticals present in aquatic environment on Phase I metabolism in fish. *Environ. Toxicol. Pharmacol.* **2015**, *40*, 430–444. [CrossRef] [PubMed]
2. Ebele, A.J.; Abdallah, M.A.-E.; Harrad, S. Pharmaceuticals and personal care products (PPCPs) in the freshwater aquatic environment. *Emerg. Contam.* **2017**, *3*, 1–6. [CrossRef]
3. Collado, N.; Rodriguez-Mozaz, S.; Gros, M.; Rubirola, A.; Barcelo, D.; Comas, J.; Rodriguez-Roda, I.; Buttiglieri, G. Pharmaceuticals occurrence in a WWTP with significant industrial contribution and its input into the river system. *Environ. Pollut.* **2014**, *185*, 202–212. [CrossRef] [PubMed]
4. Frederic, O.; Yves, P. Pharmaceuticals in hospital wastewater: Their ecotoxicity and contribution to the environmental hazard of the effluent. *Chemosphere* **2014**, *115*, 31–39. [CrossRef] [PubMed]
5. Golovko, O.; Kumar, V.; Fedorova, G.; Randak, T.; Grabic, R. Removal and seasonal variability of selected analgesics/anti-inflammatory, anti-hypertensive/cardiovascular pharmaceuticals and UV filters in wastewater treatment plant. *Environ. Sci. Pollut. Res.* **2014**, *21*, 7578–7585. [CrossRef] [PubMed]
6. Boxall, A.B.A.; Rudd, M.A.; Brooks, B.W.; Caldwell, D.J.; Choi, K.; Hickmann, S.; Innes, E.; Ostapyk, K.; Staveley, J.P.; Verslycke, T.; et al. Pharmaceuticals and Personal Care Products in the Environment: What Are the Big Questions? *Environ. Health Perspect.* **2012**, *120*, 1221–1229. [CrossRef] [PubMed]
7. Huerta, B.; Rodriguez-Mozaz, S.; Barcelo, D. Pharmaceuticals in biota in the aquatic environment: Analytical methods and environmental implications. *Anal. Bioanal. Chem.* **2012**, *404*, 2611–2624. [CrossRef]
8. Schulz, M.; Iwersen-Bergmann, S.; Andresen, H.; Schmoldt, A. Therapeutic and toxic blood concentrations of nearly 1,000 drugs and other xenobiotics. *Crit. Care* **2012**, *16*, R136. [CrossRef]
9. Fedorova, G.; Golovko, O.; Randak, T.; Grabic, R. Storage effect on the analysis of pharmaceuticals and personal care products in wastewater. *Chemosphere* **2014**, *111*, 55–60. [CrossRef]
10. Yadav, M.K.; Short, M.D.; Aryal, R.; Gerber, C.; van den Akker, B.; Saint, C.P. Occurrence of illicit drugs in water and wastewater and their removal during wastewater treatment. *Water Res.* **2017**, *124*, 713–727. [CrossRef]
11. Marti, E.; Huerta, B.; Rodriguez-Mozaz, S.; Barcelo, D.; Marce, R.; Balcazar, J.L. Abundance of antibiotic resistance genes and bacterial community composition in wild freshwater fish species. *Chemosphere* **2018**, *196*, 115–119. [CrossRef] [PubMed]
12. Stankiewicz, A.; Giebultowicz, J.; Stankiewicz, U.; Wroczynski, P.; Nalecz-Jawecki, G. Determination of selected cardiovascular active compounds in environmental aquatic samples—Methods and results, a review of global publications from the last 10 years. *Chemosphere* **2015**, *138*, 642–656. [CrossRef] [PubMed]
13. Buric, M.; Grabicova, K.; Kubec, J.; Kouba, A.; Kuklina, I.; Kozak, P.; Grabic, R.; Randak, T. Environmentally relevant concentrations of tramadol and citalopram alter behaviour of an aquatic invertebrate. *Aquat. Toxicol.* **2018**, *200*, 226–232. [CrossRef] [PubMed]
14. Fong, P.P.; Ford, A.T. The biological effects of antidepressants on the molluscs and crustaceans: A review. *Aquat. Toxicol.* **2014**, *151*, 4–13. [CrossRef]
15. Hyttel, J. Pharmacological Characterization of Selective Serotonin Reuptake Inhibitors (SSRIs). *Int. Clin. Psychopharm.* **1994**, *9*, 19–26. [CrossRef]
16. Skerritt, J.H.; Johnston, G.A. Enhancement of GABA binding by benzodiazepines and related anxiolytics. *Eur. J. Pharmacol.* **1983**, *89*, 193–198. [CrossRef]

17. Rosi-Marshall, E.J.; Snow, D.; Bartelt-Hunt, S.L.; Paspalof, A.; Tank, J.L. A review of ecological effects and environmental fate of illicit drugs in aquatic ecosystems. *J. Hazard. Mater.* **2015**, *282*, 18–25. [CrossRef] [PubMed]
18. Grabicova, K.; Grabic, R.; Blaha, M.; Kumar, V.; Cerveny, D.; Fedorova, G.; Randak, T. Presence of pharmaceuticals in benthic fauna living in a small stream affected by effluent from a municipal sewage treatment plant. *Water Res.* **2015**, *72*, 145–153. [CrossRef]
19. Klaminder, J.; Brodin, T.; Sundelin, A.; Anderson, N.J.; Fahlman, J.; Jonsson, M.; Fick, J. Long-Term Persistence of an Anxiolytic Drug (Oxazepam) in a Large Freshwater Lake. *Environ. Sci. Technol.* **2015**, *49*, 10406–10412. [CrossRef]
20. Lagesson, A.; Fahlman, J.; Brodin, T.; Fick, J.; Jonsson, M.; Bystrom, P.; Klaminder, J. Bioaccumulation of five pharmaceuticals at multiple trophic levels in an aquatic food web—Insights from a field experiment. *Sci. Total. Environ.* **2016**, *568*, 208–215. [CrossRef]
21. Brodin, T.; Piovano, S.; Fick, J.; Klaminder, J.; Heynen, M.; Jonsson, M. Ecological effects of pharmaceuticals in aquatic systems-impacts through behavioural alterations. *Philos. Trans. R. Soc. B* **2014**, *369*. [CrossRef] [PubMed]
22. Brodin, T.; Nordling, J.; Lagesson, A.; Klaminder, J.; Hellstrom, G.; Christensen, B.; Fick, J. Environmental relevant levels of a benzodiazepine (oxazepam) alters important behavioral traits in a common planktivorous fish, (*Rutilus rutilus*). *J. Toxicol. Environ. Health A* **2017**, *80*, 963–970. [CrossRef] [PubMed]
23. Pal, R.; Megharaj, M.; Kirkbride, K.P.; Naidu, R. Illicit drugs and the environment—A review. *Sci. Total Environ.* **2013**, *463*, 1079–1092. [CrossRef] [PubMed]
24. Nielsen, S.V.; Kellner, M.; Henriksen, P.G.; Olsen, H.; Hansen, S.H.; Baatrup, E. The psychoactive drug Escitalopram affects swimming behaviour and increases boldness in zebrafish (*Danio rerio*). *Ecotoxicology* **2018**, *27*, 485–497. [CrossRef] [PubMed]
25. Hossain, M.S.; Patoka, J.; Kouba, A.; Buric, M. Clonal crayfish as biological model: A review on marbled crayfish. *Biologia* **2018**, *73*, 841–855. [CrossRef]
26. Kubec, J.; Kouba, A.; Buřič, M. Communication, behaviour, and decision making in crayfish: A review. *Zool. Anz.* **2018**, *278*, 28–37. [CrossRef]
27. Barry, M.J. Effects of fluoxetine on the swimming and behavioural responses of the Arabian killifish. *Ecotoxicology* **2013**, *22*, 425–432. [CrossRef]
28. Jonsson, M.; Fick, J.; Klaminder, J.; Brodin, T. Antihistamines and aquatic insects: Bioconcentration and impacts on behavior in damselfly larvae (Zygoptera). *Sci. Total Environ.* **2014**, *472*, 108–111. [CrossRef]
29. Cunha, D.L.; Mendes, M.P.; Marques, M. Environmental risk assessment of psychoactive drugs in the aquatic environment. *Environ. Sci. Pollut. Res.* **2019**, *26*, 78–90. [CrossRef]
30. Grabic, R.; Fick, J.; Lindberg, R.H.; Fedorova, G.; Tysklind, M. Multi-residue method for trace level determination of pharmaceuticals in environmental samples using liquid chromatography coupled to triple quadrupole mass spectrometry. *Talanta* **2012**, *100*, 183–195. [CrossRef]
31. Sehonova, P.; Svobodova, Z.; Dolezelova, P.; Vosmerova, P.; Faggio, C. Effects of waterborne antidepressants on non-target animals living in the aquatic environment: A review. *Sci. Total Environ.* **2018**, *631–632*, 789–794. [CrossRef] [PubMed]
32. Courtney, D.B. Selective serotonin reuptake inhibitor and venlafaxine use in children and adolescents with major depressive disorder: A Systematic Review of Published Randomized Controlled Trials. *Can. J. Psychiatry* **2004**, *49*, 557–563. [CrossRef] [PubMed]
33. Loos, R.; Carvalho, R.; Antonio, D.C.; Cornero, S.; Locoro, G.; Tavazzi, S.; Paracchini, B.; Ghiani, M.; Lettieri, T.; Blaha, L.; et al. EU-wide monitoring survey on emerging polar organic contaminants in wastewater treatment plant effluents. *Water Res.* **2013**, *47*, 6475–6487. [CrossRef] [PubMed]
34. Grabicova, K.; Grabic, R.; Fedorova, G.; Fick, J.; Cerveny, D.; Kolarova, J.; Turek, J.; Zlabek, V.; Randak, T. Bioaccumulation of psychoactive pharmaceuticals in fish in an effluent dominated stream. *Water Res.* **2017**, *124*, 654–662. [CrossRef] [PubMed]
35. Brodin, T.; Fick, J.; Jonsson, M.; Klaminder, J. Dilute concentrations of a psychiatric drug alter behavior of fish from natural populations. *Science* **2013**, *339*, 814–815. [CrossRef] [PubMed]
36. Hossain, M.S.; Kubec, J.; Grabicova, K.; Grabic, R.; Randak, T.; Guo, W.; Kouba, A.; Buřič, M. Environmentally relevant concentrations of methamphetamine and sertraline modify the behavior and life history traits of an aquatic invertebrate. *Sci. Total Environ.* **2019**, under review.

37. Gherardi, F. Behaviour. In *Biology of Freshwater Crayfish*; Holdich, D.M., Ed.; Blackwell Science: Oxford, UK, 2002; pp. 258–290.
38. Holdich, D.M. Background and functional morphology. In *Biology of Freshwater Crayfish*; Holdich, D.M., Ed.; Blackwell Science: Oxford, UK, 2002; pp. 3–29.
39. Craddock, N.; Jones, I. Genetics of bipolar disorder. *J. Med. Genet.* **1999**, *36*, 585–594. [CrossRef]
40. Manning, A.; Dawkins, M.S. *An Introduction to Animal Behaviour*, 6th ed.; Cambridge University Press: Cambridge, UK; New York, NY, USA, 2012; p. ix, 458p.
41. Schmitz, O.J. Predator diversity and trophic interactions. *Ecology* **2007**, *88*, 2415–2426. [CrossRef]
42. Duffy, J.E.; Cardinale, B.J.; France, K.E.; McIntyre, P.B.; Thebault, E.; Loreau, M. The functional role of biodiversity in ecosystems: Incorporating trophic complexity. *Ecol. Lett.* **2007**, *10*, 522–538. [CrossRef]
43. Fong, P.P.; Bury, T.B.; Dworkin-Brodsky, A.D.; Jasion, C.M.; Kell, R.C. The antidepressants venlafaxine ("Effexor") and fluoxetine ("Prozac") produce different effects on locomotion in two species of marine snail, the oyster drill (*Urosalpinx cinerea*) and the starsnail (*Lithopoma americanum*). *Mar. Environ. Res.* **2015**, *103*, 89–94. [CrossRef]
44. Bossus, M.C.; Guler, Y.Z.; Short, S.J.; Morrison, E.R.; Ford, A.T. Behavioural and transcriptional changes in the amphipod *Echinogammarus marinus* exposed to two antidepressants, fluoxetine and sertraline. *Aquat. Toxicol.* **2014**, *151*, 46–56. [CrossRef] [PubMed]
45. Imeh-Nathaniel, A.; Rincon, N.; Orfanakos, V.B.; Brechtel, L.; Wormack, L.; Richardson, E.; Huber, R.; Nathaniel, T.I. Effects of chronic cocaine, morphine and methamphetamine on the mobility, immobility and stereotyped behaviors in crayfish. *Behav. Brain Res.* **2017**, *332*, 120–125. [CrossRef] [PubMed]
46. Di Lorenzo, T.; Castaño-Sánchez, A.; Di Marzio, W.D.; García-Doncel, P.; Martínez, L.N.; Galassi, D.M.P.; Iepure, S. The role of freshwater copepods in the environmental risk assessment of caffeine and propranolol mixtures in the surface water bodies of Spain. *Chemosphere* **2019**, *220*, 227–236. [CrossRef] [PubMed]
47. Reynolds, J.; Holdich, D. Growth and reproduction. In *Biology of Freshwater Crayfish*; Holdich, D.M., Ed.; Blackwell Science: Oxford, UK, 2002; pp. 152–191.
48. Carfagno, G.L.F.; Fong, P.P. Growth Inhibition of Tadpoles Exposed to Sertraline in the Presence of Conspecifics. *J. Herpetol.* **2014**, *48*, 571–576. [CrossRef]

Review

Regulation of Perfluorooctanoic Acid (PFOA) and Perfluorooctane Sulfonic Acid (PFOS) in Drinking Water: A Comprehensive Review

Frederick Pontius

Department of Civil Engineering and Construction Management, Gordon and Jill Bourns College of Engineering, California Baptist University, 8432 Magnolia Ave., Riverside, CA 92503, USA; fpontius@calbaptist.edu

Received: 14 August 2019; Accepted: 18 September 2019; Published: 26 September 2019

Abstract: Perfluorooctanoic acid (PFOA) and perfluorooctane sulfonic acid (PFOS) are receiving global attention due to their persistence in the environment through wastewater effluent discharges and past improper industrial waste disposal. They are resistant to biological degradation and if present in wastewater are discharged into the environment. The US Environmental Protection Agency (USEPA) issued drinking water Health Advisories for PFOA and PFOS at 70 ng/L each and for the sum of the two. The need for an enforceable primary drinking water regulation under the Safe Drinking Water Act (SDWA) is currently being assessed. The USEPA faces stringent legal constraints and technical barriers to develop a primary drinking water regulation for PFOA and PFOS. This review synthesizes current knowledge providing a publicly available, comprehensive point of reference for researchers, water utilities, industry, and regulatory agencies to better understand and address cross-cutting issues associated with regulation of PFOA and PFOS contamination of drinking water.

Keywords: perfluorooctanoic acid (PFOA); perfluorooctane sulfonic acid (PFOS; drinking water; Safe Drinking Water Act (SDWA); US Environmental Protection Agency (USEPA); regulation; best available technology (BAT); maximum contaminant level (MCL)

---

1. Introduction

Perfluoroalkyl substances (PFASs) are synthetic industrial chemicals used throughout the world. The Swedish Chemicals Agency (2015) [1] estimated that more than 4000 types of perfluoroalkyl substances have been synthesized, with more than 2000 on the global market. Global annual PFAS emissions steadily increased from 2002 to 2012, with a geographical shift of industrial sources away from North America, Europe and Japan towards emerging Asian economies and China [2,3]. The Organization for Economic Cooperation and Development (OECD) has updated a comprehensive list of over 4700 PFAS-related chemicals on the global market based on the chemical abstract service (CAS) numbers [4].

PFASs are resistant to biological degradation, breaking down very slowly in the environment. When present in municipal and industrial wastewaters they typically pass through wastewater treatment plant processes and are discharged in the effluent. When used industrially, PFASs may be emitted into the air or accidently released to the environment [2,3]. They are poorly adsorbed by soils and aquifer materials and readily transported via surface water and ground water. Because of their longevity in the environment, PFASs and other synthetic organic chemicals with these properties are generally referred to as persistent organic pollutants (POPs).

To date most research attention globally has been given to two PFASs, perfluorooctanoic acid (PFOA) and perfluorooctane sulfonic acid (PFOS). Potential adverse human health effects from PFOA and PFOS in drinking water and POPs in the environment have been an ongoing concern for several decades. Improper disposal of industrial wastes containing PFASs in the decades prior to the enactment of environmental protection laws, PFAS presence in wastewater discharges, and releases to the air have distributed PFAS contaminants globally. Regulatory agencies in more than 12 countries have established guidelines or health advisory values for PFOA and PFOS in drinking water and/or groundwater.

In 2009 the US Environmental Protection Agency (USEPA) identified PFOA and PFOS as contaminants of potential concern in drinking water. Major PFOA and PFOS contamination of the environment and drinking water supplies had been discovered in several states, in particular West Virginia [5], Ohio [6], and Minnesota [7]. PFOA and PFOS were included on the third drinking water contaminant candidate list (CCL) [8]. The CCL lists substances of potential health concern if present in drinking water, but for which the knowledge-base is too limited to determine whether a national regulation is needed. The Safe Drinking Water Act (SDWA) requires updating of the CCL every five years. Where drinking water is contaminated with PFOA and PFOS, state agencies take action to address these situations.

Drinking water health advisories for PFOA and PFOS were issued by the USEPA in May 2016 [9,10]. Advisories provide technical information to state agencies and other public health officials on health effects, analytical methodologies, and treatment technologies to assist them in making risk management decisions. Advocacy groups and others were critical of the advisories, arguing that PFOA and PFOS should be regulated nationally [11,12]. In response, on February 14, 2019 the USEPA released a comprehensive action plan to address PFASs [13,14], stating for drinking water that:

"The next step in the Safe Drinking Water Act (SDWA) process for issuing drinking water standards is to propose a regulatory determination. The Agency is also gathering and evaluating information to determine if regulation is appropriate for a broader class of PFAS".

State agencies have the authority to act to address PFOA and PFOS contamination within their jurisdiction. States may adopt health advisories and regulations more stringent than the USEPA. Issuance of a national enforceable regulation for PFOA and PFOS involves considerations beyond the state level. PFOA and PFOS have drawn worldwide research attention, so much so that the knowledge-base for these contaminants has expanded dramatically since 2016. The USEPA is now deciding whether national drinking water regulations for PFOA and PFOS are warranted.

Originally enacted in 1974, the SDWA was amended in 1996 to establish specific procedures and conditions for issuing an enforceable national U.S. drinking water regulation [15]. Regulation of a contaminant in drinking water raises cross-cutting issues related to health, analysis, exposure, water treatment, risk assessment, and risk management. The purpose of this paper is to synthesize current knowledge to serve as a publicly available comprehensive starting point for researchers, water utilities, industry, and regulatory agencies to understand the issues most relevant to a drinking water regulation for PFOA and PFOS. This paper examines the rationale behind SDWA requirements for contaminant regulation as they apply to PFOA, PFOS, and other POPs that may be regulated in the future. Current knowledge regarding PFOA and PFOS is summarized and recommendations made for responding to PFOA and PFOS contamination regardless of whether a national regulation is established.

2. Authority to Regulate

The SDWA regulatory process is very thorough, taking several years to complete. The SDWA, the Administrative Procedure Act (APA) [16], and agency regulatory policies impose important constraints on contaminant regulation.

The majority of drinking water contaminants now regulated were established between 1986 and 1995 with many maximum contaminant levels (MCLs) based on prior regulations. In 1986 the SDWA was amended to require regulation of at least 25 contaminants every five years. Lack of information seriously hindered the regulation of new contaminants. Resource limitations made it impractical for the agency to meet this requirement, and regulatory activity came to a halt in the mid-1990s. To meet the SDWA requirement initially regulations for disinfectants, disinfection byproducts and additional surface water treatment rules were developed through formal regulatory negotiation. However, regulating 25 individual contaminants on a regular basis proved to be an impossible task.

In 1990 the USEPA Science Advisory Board issued the report Reducing Risk: Setting Priorities and Strategies for Environmental Protection [17]. This report was very influential during the decade of the 1990s for focusing the USEPA on setting priorities to address environmental risks that would achieve the greatest risk reduction. In 1996 the US Congress amended the SDWA to establish procedures aimed at enabling the USEPA to set priorities for regulating contaminants in drinking water.

The SDWA grants the USEPA the general authority to regulate a contaminant in drinking water if it finds the following three conditions are met [18]:

1. The contaminant may have an adverse effect on the health of persons;
2. The contaminant is known to occur or there is a substantial likelihood that the contaminant will occur in public water systems with a frequency and at levels of public health concern; and
3. The regulation of the contaminant presents meaningful opportunity for health risk reduction for persons served by public water systems.

The SDWA conditions to regulate a new contaminant are intended to focus the agency's efforts and limited resources on regulating contaminants which would achieve the greatest risk reduction nationally. Substances not present or are not likely to occur in drinking water nationally pose no actual health risk. A substance that does not and is not likely to have an adverse health effect poses no actual health risk. Drinking water regulations are to achieve a meaningful health risk reduction nationally. The SDWA also allows the agency to act to address contaminants that pose an urgent threat to public health [19].

3. Properties and Uses of PFOA and PFOS

PFOA and PFOS are both considered long-chain PFASs, which are perfluorocarboxilic acids with eight or more carbon atoms or perfluorosulfonic acids with six or more carbon atoms [20]. PFOA and PFOS are eight-carbon compounds (C8) having unique chemical properties including surface activity, thermal and acid resistance, and repelling both water and oil [20]. Commercial applications of PFASs include stain-resistant coatings for carpeting and upholstery, breathable water-resistant outdoor clothing, and greaseproof packaging. They are also used to manufacture fluoropolymers such as polytetrafluoroethylene [21]. PFASs are found in aqueous film-forming foams used to fight hydrocarbon fires [22].

PFOA was first used to manufacture commercial products in 1949 and was used to manufacture polytetrafluoroethylene (PTFE) for non-stick coatings. PFOS had been produced since the 1940s and was previously used in fabric protectors. Both PFOA and PFOS were used in a variety of other industrial and well-known consumer products. PFOA and PFOS are stable, non-volatile, and very soluble in water. They are highly mobile in the environment and have been detected in natural waters, wastewater effluents, treated drinking waters and a variety of food products in many countries [23,24]. When inhaled or ingested, PFOA and PFOS are readily absorbed into the human body. In 2009–2010 the National average blood serum levels in the United States were 3.1 ppb and 9.3 ppb for PFOS and PFOA, respectively [25]. These compounds are biologically stable and are not metabolized.

Major US manufacturers have voluntarily phased out production of PFOA and PFOS. In the year 2000 the US manufacturer 3M announced it would voluntarily phase out all production of PFOS due to concerns over potential lawsuits, regulatory pressure, and negative public perception. In 2006, eight US chemical manufacturers agreed to phase out all production and use of PFOA and related compounds by 2015. Both PFOA and PFOS are no longer produced in the United States. Since elimination of their use, blood serum levels for PFOA and PFOS in the United States have been declining [26]. Exposure to PFOA and PFOS remains possible due to legacy uses, existing and legacy uses on imported goods, degradation of precursors, and extremely high persistence in the human body and the environment. Although their use has been discontinued in the United States, they are still produced for commercial use in other countries [27,28].

4. Analytical Methods

Aqueous samples are typically analyzed for PFASs using liquid-liquid extraction, ion-pair extraction, or solid-phase extraction followed by HPLC-MS/MS or GC/MS [29]. In 2009, the USEPA released Method 537 for the analysis of 14 PFASs in drinking water using solid-phase extraction followed by Liquid Chromatography/Tandem Mass Spectrometry (LC/MS/MS) [30]. A highly-skilled analyst is required to generate reliable analytical results using this method. For US drinking water regulatory reporting, a laboratory must use an analytical method developed by the USEPA or a USEPA-approved equivalent method. Method 537 had limitations but garnered widespread application for analyzing water samples for PFOA and PFOS. The 2009 version of the method was expanded and improved to include additional analytes and lower detection limits.

4.1. USEPA Method 537.1

Method 537.1 [31] is the standard method for the analysis of 18 PFASs, and was issued in November 2018. Table 1 lists each PFAS covered by Method 537.1, with each acronym, Chemical Abstract Service Registration Number (CASRN), Detection Limit (DL), and single laboratory Lowest Concentration Minimum Reporting Level (LCMRL). The DL characterizes the accuracy of each method and is defined as the statistically calculated minimum concentration that can be measured with 99% confidence that the reported value is greater than zero. The DL is compound-dependent and is affected by extraction efficiency, sample matrix, fortification concentration, and instrument performance. The LCMRL is the lowest true concentration for which future recovery is predicted to fall, with high confidence (99%), and between 50 and 150% recovery.

Table 1. Compounds detectable by USEPA Method 537.1 [31]. USEPA: US Environmental Protection Agency; CASRN: Chemical Abstract Service Registration Number; LCMRL: Lowest Concentration Minimum Reporting Level; DL: Detection Limit.

Analyte (Chain Length)	Acronym	CASRN	DL ng/L	LCMRL ng/L
Hexafluoropropylene oxide dimer acid	HFPO-DA	13252-13-6	1.9	4.3
N-ethyl perfluoroactanesulfonamido-acetic acid	NEtFOSAA	2991-50-6	2.8	4.8
N-methyl perfluorooctanesulfonamidoacetic acid	NMeFOSAA	2355-31-9	2.4	4.3
Perfluorobutanesulfonic acid (C4)	PFBS	375-73-5	1.8	6.3
Perfluorodecanoic acid (C10)	PFDA	335-76-2	1.6	3.3
Perfluorododecanoic acid (C12)	PFDoA	307-55-1	1.2	1.3
Perfluoroheptanoic acid (C7)	PFHpA	375-85-9	0.71	0.63
Perfluorohexanesulfonic acid (C6)	PFHxS	355-46-4	1.4	2.4
Perfluorohexanoic acid (C6)	PFHxA	307-24-4	1.0	1.7
Perfluorononanoic acid (C9)	PFNA	375-95-1	0.70	0.83
Perfluorooctanesulfonic acid (C8)	PFOS	1763-23-1	1.1	2.7
Perfluorooctanoic acid (C8)	PFOA	335-67-1	0.53	0.82
Perfluorotetradecanoic acid (C14)	PFTA	376-06-7	1.1	1.2
Perfluorotridecanoic acid (C13)	PFTrDA	72629-94-8	0.72	0.53
Perfluoroundedcanoic acid (C11)	PFUnA	2058-94-8	1.6	5.2
11-chloroeicosafluoro-3-oxaundecane-1-sulfonic acid	11Cl-PF3OUdS	763051-92-9	1.5	1.5
9-chlorohexadecafluoro-3-oxanone-1-sulfonic acid	9Cl-PF3ONS	756426-58-1	1.4	1.8
4,8-dioxa-3H-perfluorononanoic acid	ADONA	919005-14-4	0.88	0.55

4.2. Unregulated Contaminant Monitoring

In the United States, PFOA and PFOS are two of 97 chemicals listed on the fourth drinking water contaminant candidate list (CCL 4) [32] published in 2016. As noted above, for a listed substance to be regulated, sufficient knowledge must be available in order for a regulation to meet the SDWA regulatory requirements.

To assess the extent of contamination in the United States, the SDWA authorizes collection of nationwide occurrence data through the unregulated contaminant monitoring rule (UCMR) [33]. Under the UCMR, occurrence data is collected for a maximum of 30 analytes in a five-year cycle. Samples are collected at all public water systems serving >10,000 people and at a statistical sample of public water systems serving <10,000 people. Monitoring of PFOA, PFOS, perfluorobutanesulfonic acid (PFBS), perfluoroheptanoic acid (PFHpA), perfluorohexanesulfonic acid (PFHxS), and perfluorononanoic acid (PFNA) was required in the third UCMR (UCMR 3) between 2013 and 2015 [34].

4.3. Minimum Reporting Level (MRL)

For each contaminant monitored under the UCMR, a minimum reporting level (MRL) is specified. The MRL is the minimum concentration at which a contaminant is reliably quantitated by individual laboratories [34]. At or above the MRL, a competent drinking water laboratory should be expected to obtain 50–150% recovery or better. The MRL differs from the minimum detection level by considering both the standard deviation of low concentration analyses (precision) and the accuracy of the measurements as they impact achievement of data quality objectives for spike recovery. MRLs reflect the performance of competent commercial laboratories and are not based on the performance of a particular instrument or single laboratory. Table 1 presents the MRLs for the PFASs monitored under UCMR3.

4.4. Practical Quantitation Level (PQL)

For a drinking water regulation to be enforceable the MCL must be set at a concentration providing a clear delineation between test results above and below the standard not affected by variability in laboratory performance. The practical quantitation level (PQL) is typically used for this purpose. The PQL is the lowest concentration of an analyte that can be reliably measured within specified limits of precision and accuracy during routine laboratory operating conditions. The PQL, MRL, DL, and LCMRL for a contaminant may all differ. An interim reporting limit and PQL for PFOA of 6 ng/L has been adopted by New Jersey to support development of groundwater quality standards [35]. A PQL for PFOA and PFOS has not been established by the USEPA.

5. Occurrence in Public Water Systems

Drinking water supplies are vulnerable to PFOA and PFOS contamination from a variety of sources. PFOA was first discovered because of harmful effects due to leakage from landfills that had received PFAS-related industrial wastes [5,6,36–38]. Airports and fire training areas have been contaminated by PFASs contained in aqueous film-forming foams used during firefighting training activities [39]. Wastewater treatment plant discharges, biodegradation of precursors during wastewater treatment, and land application of biosolids are potential sources of PFOA and PFOS in drinking water supplies. Once in the environment PFOA and PFOS are stable and bioaccumulate into the food chain [40,41].

PFOA, PFOS, and many PFASs have been detected in the environment and in drinking water worldwide. Selected studies and the range of PFOA and PFOS concentrations reported are listed in Table 2. Most studies listed analyzed samples for several PFASs in addition to PFOA and PFOS. A few studies reported the ΣPFAS with the number of compounds tested differing between studies. The concentrations presented in Table 2 are not directly comparable between studies because of differences in the number of compounds tested, the analytical methods used, and the prevailing laboratory quality assurance and quality control (QA/QC).

Table 2. Concentrations of perfluorooctanoic acid (PFOA) and perfluorooctane sulfonic acid (PFOS) reported in selected studies.

Country	Location	PFOA ng/L	PFOS ng/L	ΣPFAS ng/L	References
Australia	Drinking water	nd–9.7	nd–16	nd–28	[42]
	Recycled water	<0.09–6.9	<0.03–34	<0.03–74	[43]
	Sydney Harbor	4.2–6.4	7.5–21		[44]
Canada	Great Lakes	1.6–6.7	1.2–37.6		[45]
	Lake Ontario tributaries	4.1–38.1	2.6–22.9		[45]
	WWTP effluent	6.5–54.7	8.6–208.5	10 max	[45]
	River waters	0.8 max	2.5 max	44 max	[24]
	Drinking water	0.20–2.1	3.3	10	[46]
	Drinking water	nd–4.86	nd–4.99		[47]
	Bottled water	nd–<0.2	nd–<0.1		[47]
China	Huangpu River	1590	20.5		[48]
	Pearl River tributaries	0.85–13	0.90–99		[49]
	Yangtze River	2.1–260	<0.01–14		[49]
	WWTP influent	<0.13–20	<0.13–10	11.1–80.6	[50]
	WWTP influent	2–91	1–32		[51]
	WWTP effluent	<0.13–20	<0.13–16	4.8–71.5	[50]
	WWTP effluent	3–107	1–67		[51]
	Drinking water	nd–26.3	0.01–2.80	4.49–174.93	[52]
	Drinking water	10 mean	3.9 mean	180 max	[46]
	Drinking water	<0.1–45.9	<0.1–14.8		[53]
	Bottled water	nd–0.95	nd		[47]
Germany	River Elbe	7.2–9.6	0.5–2.9		[54]
	WWTP effluent	7.6–12.3	<0.06–82.2		[54]
	Drinking water	0.50	0.69		[46]
India	Surface waters	4–93	3–29		[55]
	Drinking water	<0.033–2.0	<0.04–8.4		[46]
Japan	River waters	0.7–43,239	nd–200		[24]
	SW sources	5.2–92	0.26–22		[56]
	Drinking water	2.3–84	0.16–22		[56]
	Drinking water	0.18–18	0.066–4.9		[46]
The Netherlands	Drinking water	<0.6–4.9	<0.6–3.0	<0.6–54	[57]
	GW near a fluoro-polymer plant	3900–25,000			[58]
Taiwan	Semiconductor WWTP effluent	118.3	128,670		[59]
	River waters	10.9–310	82–5400		[59]
	Drinking water	3.7	5.4		[46]
Thailand	River waters	<0.3–450	21 max		[24]
	Drinking water	1.2–4.6	0.13–1.9		[46]
United States	SW sources	112 max	48.3 max	0.12–1101	[60]
	Treated DW	104 max	36.9 max	0.15–1094	[60]
	Drinking water	1.2	1.4		[46]
	Drinking water	<5–30	<1–57		[61]

nd = none detected; WWTP = wastewater treatment plant; SW = surface water; GW = groundwater.

The number of water systems impacted is a key factor when determining whether to regulate a contaminant under the SDWA. As mentioned above, a contaminant must be known or likely to be present in drinking water at levels of public health concern to warrant a national regulation. Table 3 presents the results of PFAS monitoring under UCMR 3 [33]. MRLs for PFOA and PFOS were set at 20 ng/L and 40 ng/L, respectively. At least one of the PFASs was detected in samples taken from 36 states and territories. MRLs do not represent levels considered "significant" or "harmful." Detection of a contaminant above the MRL does not necessarily represent cause for concern [34]. All six PFASs were tested at 4920 public water systems. The MRL for PFOA was exceeded by 117 water systems representing 2.4% of the water systems tested. The MRL for PFOS was exceeded in 95 water systems, representing 2.2% of the water systems tested.

Table 3. PFAS monitoring results under the unregulated contaminant monitoring rule (UCMR) 3 [34].[1]

Measure	PFOA	PFOS	PFNA	PFHxS	PFHpA	PFBS
MRL ng/L	20	40	20	30	10	90
HRL ng/L	70	70	na	na	na	na
Total number of test results	36,972	36,972	36,972	36,971	36,972	36,972
Number of results ≥ MRL	379	292	19	207	236	19
Number of results > HRL	32	124	na	na	na	na
% of all results > HRL	0.09%	0.34%	na	na	na	na
Number of PWSs with results	4920	4920	4920	4920	4920	4920
Number of PWSs > HRL	13	46	na	na	na	na
% of PWSs > HRL	0.3%	0.9%	na	na	na	na

[1] MRL: Minimum Reporting Level, HRL: Health Reference Level, PWSs: public water systems.

The UCMR 3 monitoring results represent the only available statistically valid survey of national occurrence for the PFASs tested. When analyzing UCMR data the numbers of water systems exceeding the MRL are a function of the concentration at which the MRL is established. Water systems impacted by PFASs are unevenly distributed across the United States, and the number of systems affected may be greater than suggested by UCMR3 data [62]. One laboratory performing UCMR 3 testing reanalyzed its own UCMR 3 data (~1800 water systems) using an in-house MRL at 5 ng/L for all six PFAS. The percentage of water systems detecting at least one of the six PFASs was 5.3%, as compared to the USEPA estimate of 3.9% at the higher MRLs [62].

To assess the health significance of UCMR monitoring results a Health Reference Level (HRL) was established at 70 ng/L each for PFOA and PFOS [34]. HRLs are risk-derived concentrations against which occurrence data are compared to determine if contaminants may occur at levels of public health concern. HRLs do not represent final determinations but are derived as screening levels prior to development of a formal exposure assessment. The 0.07 μg/L HRL for PFOA was exceeded by 13 water systems or 0.3% of the water systems tested. The 0.07 μg/L HRL for PFOS was exceeded by 46 water systems or 0.9% of the water systems tested. At the time of UCMR 3 sampling, the USEPA provisional drinking water health advisory levels for PFOA and PFOS were 0.400 μg/L and 0.200 μg/L, respectively [63].

In a joint effort between the USEPA and the U.S. Geological Survey (USGS), paired samples from 25 drinking water treatment plants were analyzed for 247 chemical and microbial contaminants of emerging concern. Data from this study on the occurrence of PFSA in the source water are now available [60]. Although the number of water plants sampled do not represent a statistically valid national estimate, the results do provide insight into the occurrence of PFOA and PFOS. PFOA was detected in 100% of source water samples, was quantified in 76% of samples, and occurred at a median concentration of 6.32 μg/L with a maximum of 112 μg/L. PFOS was detected in 96% of source water samples, was quantified in 88% of samples, and occurred at a median concentration of 0.00228 μg/L with a maximum of 0.0483 μg/L.

Of the 17 PFASs monitored in the USEPA/USGS study, 14 were qualitatively detected and 12 were quantitatively detected at least once in source water samples [60]. PFOA and PFBS were the only analytes qualitatively detected in the source water samples at all 25 water plants. PFOS, PFBA, PFDA, PFHpA, PFHxS, PFHxA, PFNA, and PFPeA were qualitatively detected in a least 90% of source water samples.

When selecting contaminants in drinking water to regulate, the USEPA is to select contaminants presenting the greatest public health concern. Effects of contaminants upon subgroups of the general population at greater risks are also considered discussed below.

6. Human Health Effects

Considerable research attention has been given identifying the effect of PFOA and PFOS exposure on human health. The PFOA and PFOS health advisory level of 0.07 μg/L for each and the sum of the two are based on assessment of the health effects information available at that time. Health effects assessments are performed in accordance with USEPA guidelines for human health risk assessment [64]. Adverse effects observed following exposure to PFOA and PFOS are the same or similar [10] and include effects in humans on serum lipids, birth weight, and serum antibodies [65,66]. A few animal studies observed effects on the liver, neonate development, and responses to immunological challenges. Both compounds were associated with tumors in long-term animal studies [67]. Detailed reviews of the toxicological and epidemiological issues associated with PFOA and PFOS have been published [68,69]. Key risk assessment decisions affecting the regulation of PFOA and PFOS are discussed below.

6.1. Exposure Assessment

Assessing potential human health risks from exposure to a contaminant in drinking water requires professional judgment to evaluate the applicability of both animal and human studies to drinking water exposures. PFOA and PFOS are excreted from the body very slowly [70]. The half-life of PFOA in the human body has been estimated in the range of 2.3 to 3.94 years and 4.3 years for PFOS, respectively [71]. Biomonitoring of blood serum levels is used to assess PFOA and PFOS exposure and body burden. Measured serum levels are then analyzed using a single-compartment pharmacokinetic model to estimate a corresponding drinking water concentration [72].

Blood serum levels are highly dependent the amount of PFOA and PFOS taken into the body. Drinking water is a major source of exposure contributing to increased serum levels [73]. Figure 1 illustrates the effect of PFOA and PFOS exposure on blood serum levels from several studies in different settings [26]. Occupational exposures (e.g., 3M workers and Dupont workers) result in the highest serum levels. Serum levels at contaminated sites depend upon drinking water concentration, age (e.g., Pease NH community), length of time since exposure was discontinued (e.g., Northern Alabama Community), and whether water treatment was installed. Blood serum levels in the general population (e.g., NHANES) now are much lower than those at prior contaminated sites shown in Figure 1.

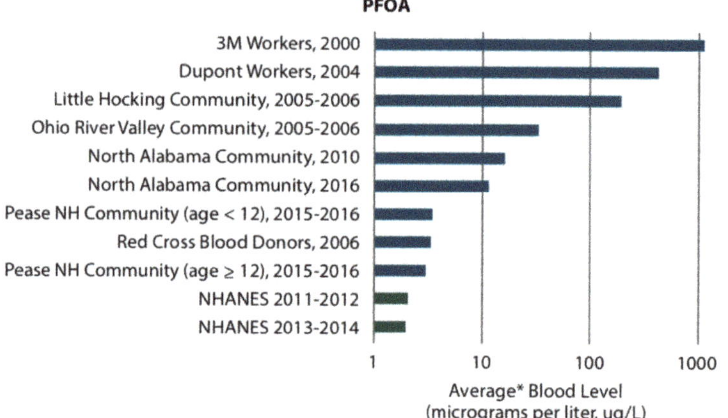

Figure 1. *Cont.*

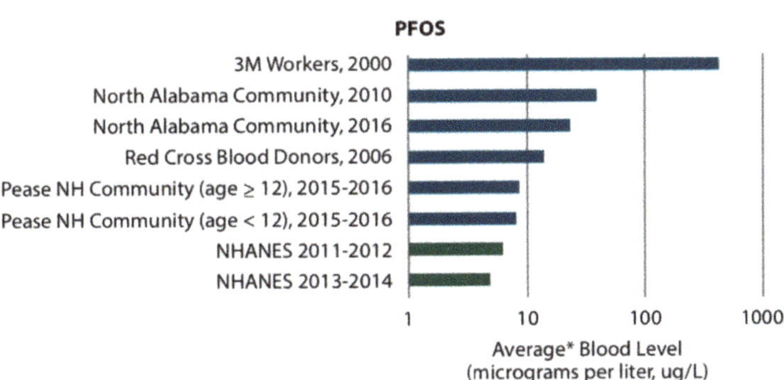

Figure 1. Blood levels in people who were exposed to perfluorooctanoic acid (PFOA, upper panel) and perfluorooctane sulfonic acid (PFOS, lower panel) [26]. Figure 1 references: 3M Workers, 2000 [74]; Dupont Workers, 2004 [75]; Little Hocking Community, 2005-2006 [76]; Ohio River Valley Community, 2005-2006 [77]; North Alabama Community [78]; Pease NH Community [79]; Red Cross Blood Donors, 2006 [80]; NHANES [25].

6.2. *Reference Dose*

The Reference Dose (RfD) represents an estimate (with uncertainty spanning perhaps an order of magnitude) of a daily human exposure to the human population (including sensitive subgroups) that is likely to be without an appreciable risk of deleterious effects during a lifetime [81]. The RfD considers adverse health effects other than cancer calculated in accordance with USEPA guidelines [81].

The health advisory RfD for PFOA is based on a pharmacokinetic Human Equivalent Dose (HED) derived from serum levels at the lowest observed adverse effects level (LOAEL) for a developmental study in mice [82]. An uncertainty factor of 300 was applied to account for "extrapolation from a LOAEL to a no observed adverse effect level (NOAEL), variability in the human population, and differences in the ways humans and rodents respond to the PFOA that reaches their tissues" [9]. Human studies demonstrate an association of PFOA exposure and effects on serum lipids, antibody responses, fetal growth and development, and the liver. Human epidemiology studies were deemed sufficient to conclude that PFOA exposure is a human health hazard but were considered inadequate for quantitative risk assessment [9]. An RfD for PFOA of 0.00002 mg/kg/day was derived based on developmental effects in neonates to provide protection to both the sensitive life stages and the general population [9].

The RfD for PFOS is based on a pharmacokinetic HED derived from serum levels at the NOAEL from a developmental study in rats [83]. An uncertainty factor of 30 was applied to account for variability in the human population and differences in human response to the PFOS reaching their tissues compared to rats [10]. An RfD of 0.00002 mg/kg/day was derived for PFOS based on the most sensitive end point to provide protection to the general population and sensitive life stages [10].

The RfD is calculated to be protective of sensitive populations. Following the 2016 issuance of USEPA health advisories the Minnesota Department of Health (MDH) reassessed its health-based guideline value for PFOA and PFOS. An Excel-based toxicokinetic model was constructed and applied to derive a guidance value for PFOA. The model incorporates body burden at birth (placental transfer), ingestion of breastmilk, and age-specific water intake rates [7]. At a relative source contribution of 50%, the calculated serum concentration allocated or 'allowed' to result from ingestion of water was 0.065 mg/L. The water concentration calculated to maintain a PFOA serum concentration at or below 0.065 mg/L throughout life for the formula-fed reasonable maximally exposed (RME) scenario and the breast-fed RME scenario was 0.15 μg/L and 0.035 μg/L, respectively [7]. Based on this assessment MDH set its final health-based PFOA guidance value at 0.035 μg/L.

6.3. Carcinogenicity

Epidemiology studies demonstrate an association of serum PFOA with kidney and testicular tumors among highly exposed members of the general population [66,67,84]. PFOA has been found to cause tumors in one or more organs of rats, including the liver, testes, and pancreas. Cancer risk was assessed following USEPA guidelines for carcinogen risk assessment [85]. Based on the weight of evidence PFOA was classified as having Suggestive Evidence of Carcinogenic Potential [9]. A quantitative dose–response assessment is not usually developed when there is only suggestive evidence unless a well-conducted study is available. A cancer dose–response based on Butenhoff et al. 2012 [86] was developed for PFOA and testicular tumors. The cancer slope factor was estimated at 0.07 mg/kg-day. A PFOA concentration of 0.5 μg/L results in a theoretical 1:1,000,000 cancer risk for an 80 kg adult drinking 2.5 L of water per day [9].

Epidemiology studies have not found a direct correlation between PFOS exposure and the incidence of carcinogenicity in humans [10]. In the only chronic oral toxicity and carcinogenicity study of PFOS in rats, liver and thyroid tumors were identified in both the controls and exposed animals at levels that did not show a direct relationship to dose [67]. The evidence for cancer in animals was judged too limited to support a quantitative cancer assessment. Based on the weight of evidence PFOS was classified as having Suggestive Evidence of Carcinogenic Potential [10]. An independent review of the carcinogenic risk of PFOS following the International Agency for Research on Cancer (IARC) evaluation process concluded "that cancer risk for PFOS, according to the IARC method, is not classifiable as carcinogenic to humans [87].

6.4. Relative Source Contribution (RSC)

The dominant source of human exposure to PFOA and PFOS is diet (water and food). Zhang et al. 2019 [73] detected both PFOA and PFOS in more than 70% of both blood samples and water samples from 13 cities in China. A correlation between the geometric mean blood levels in the general population and the corresponding mean drinking water concentration was found for PFOA ($r = 0.87$, $n = 13$, $p < 0.001$) but not observed for PFOS [73].

Paper with treated coatings have a high potential for migration of PFASs to food [88]. Food can also become contaminated with PFOA from preparation in nonstick cookware coated with polytetrafluoroethylene (PTFE) [88]. PFOA was previously used in the manufacture of several types of food packaging. In January 2016, the US Food and Drug Administration amended its food additive regulations to no longer allow for the use of PFA-containing food-contact substances [89]. Because of its widespread use in carpets, upholstered furniture, and other textiles, PFOA has been detected in indoor dust from homes, offices, vehicles, and other indoor spaces [90–92]. Workers in shops selling clothing treated with a fabric protector may also be exposed to PFASs [93]. PFOA may penetrate human skin under certain conditions [94].

Several studies have examined the relative contribution of different routes of exposure to PFOA and PFOS [92,94,95]. A Relative Source Contribution (RSC) is applied in the health advisory calculation to ensure an individual's total exposure from a contaminant does not exceed the RfD. The RSC is the portion of the RfD attributed to drinking water. The remainder to the RfD is allotted to other potential sources. The USEPA followed an Exposure Decision Tree methodology to derive an RSC of 20% for PFOA and PFOS each. The RSC for the health advisory is based on exposure to the general population. In cases where data are lacking an RSC value of 20% is used as a minimum default value. If exposure to sources other than drinking water are not expected then the RSC may be raised to 80% maximum default value.

6.5. Immunosuppression

Both human and animal studies have demonstrated the potential effect of PFOA and PFOS on the immune systems. In 2016 the National Toxicology Program (NTP) released a monograph on immunotoxicity associated with exposure to PFOA or PFOS [96]. The goal of the study was to assess evidence of PFOA or PFOS being associated with immunotoxicity in humans. Their literature search and screening process identified 33 human studies, 93 animal studies, and 27 in vitro/mechanistic studies relevant to this assessment. The NTP concluded that both PFOA and PFOS are presumed to be an immune hazard to humans based on a high level of evidence that they suppress the antibody response from animal studies and a moderate level of evidence from studies in humans [96]. The report acknowledges that the mechanisms of toxicity for both compounds are not clearly understood.

Several studies have identified immunotoxicity as an important effect of PFOS [69,97,98]. The plaque-forming cell (PFC) response, which reflects suppression of the immune response to a foreign antigen, is "among the most sensitive effects" [10]. The USEPA considered but did not use immunotoxicity as the endpoint for deriving the RfD for PFOS citing a "lack of human dosing information and lack of low-dose confirmation of effects in animals for the short-duration study." [10] After reviewing the available epidemiologic studies, the agency stated [10]:

"A limitation of epidemiology studies that evaluate the immune response following PFOS exposure is that these studies have not demonstrated whether immune parameters measured in clinically normal individuals accurately reflect the risk of future immunological diseases. Given the immune system's capacity for repair and regeneration, apparent abnormalities that are detected at one point in time might resolve before producing any adverse clinical health effect".

Some investigators have suggested the RfD for PFOS (0.00002 mg/kg/day) is too high because it is based on a developmental endpoint rather than immunotoxicity [98,99]. Pachkowski et al. 2019 [100] derived a PFOS reference dose (RfD_{Im}) using decreased plaque-forming cell (PFC) response in mice as the immune endpoint. This endpoint reflects suppression of the immune response to a foreign antigen. An RfD_{Im} of 1.8 ng/kg/day was derived based on a PFOS target human serum level of 22.5 ng/mL. This target concentration was derived from the mouse NOAEL in Dong et al. (2009) [100]. An uncertainty factor of 3 was applied to "account for potential toxicodynamic differences between mice and humans". An uncertainty factor of 10 was applied as the "standard default assumption" to account for the range of sensitivity within the human population. These uncertainty factors values applied by Pachkowski et al. 2019 [100] are identical to those applied by the USEPA to derive the PFOS health advisory. The resulting estimated RfD_{Im} is approximately one-order of magnitude lower than the RfD derived by the USEPA but it is within the general range of uncertainty attributed to an RfD estimate [81].

Chang et al. [101] systematically reviewed 24 PFOA and PFOS epidemiology studies of the general population, occupationally exposed workers, children and adults. Studies reviewed included ten studies of immune biomarker levels or gene expression patterns, ten studies of atopic of allergic disorders, five studies of infectious diseases, four studies of vaccine responses, and five studies of chronic inflammatory or autoimmune conditions. The mode of action, the level, duration, and/or timing of exposure are uncertain. The investigators concluded [101]:

"With few, often methodologically limited studies of any particular health condition, generally inconsistent results, and an inability to exclude confounding bias, or chance as an explanation for observed associations, the available epidemiologic evidence is insufficient to reach a conclusion about a causal relationship between exposure to PFOA and PFOS and any immune-related health condition in humans. When interpreting such studies, an immunodeficiency should not be presumed to exist when there is no evidence of a clinical abnormality".

6.6. Maximum Contaminant Level Goal (MCLG)

A drinking water regulation for PFOA and PFOS will include a maximum contaminant level goal (MCLG) for each contaminant. The MCLG is a non-enforceable health goal which directly influences the level at which an enforceable MCL would be established, as discussed below. The USEPA's policy is to set the MCLG at zero for contaminants known to be or are probable human carcinogens. Because the evidence of carcinogenic potential for PFOA and PFOS is only suggestive, the MCLG would be calculated based on noncancer effects using the RfD and RSC.

7. Best Available Technology (BAT)

The SDWA requires the MCL to be set as close to the MCLG as is feasible when a contaminant is regulated. "Feasible" means feasible with the use of the best technology, treatment techniques, and other means which the USEPA finds after examination for efficacy under field conditions and not solely under laboratory conditions (taking cost into consideration) [102]. Technologies meeting this feasibility criterion are called best available technologies (BAT) and must be listed in each proposed and final regulation.

PFASs are not biodegradable under typical water treatment conditions and must be removed from water using physical/chemical processes. Removal of PFASs from drinking water and wastewater in bench-scale studies and with various conventional processes has been reviewed previously [103–105]. A critical review of published data on removal of PFOA and PFOS at full-scale water treatment plants is also available [106]. Technology for drinking water treatment involves a subset of a wider variety of technologies available for remediating contaminated groundwater [107,108].

7.1. Conventional Treatment

Conventional coagulation, flocculation, sedimentation, and filtration are relatively ineffective for removing PFOA and PFOS [105,106,109,110]. At coagulant doses typically applied to treat surface water coagulation removed less than 35% of PFOA and PFOS [111,112]. Bench studies of coagulation with ferric chloride and powdered activated carbon (PAC) increased removal of PFOA and PFOS up to >90% but the initial contaminant concentration was 1 mg/L [112]. Polyaluminum chloride at an initial dose of 5 mg/L was found more effective than alum and ferric for removing PFOA and PFOS [113]. Polyaluminum chloride enhanced with PAC addition was found effective for removing PFOA [114].

A natural coagulant (*Moringe oleifera*) proved to be very effective in removing PFOA and PFOS compared to conventional coagulants, with reduction efficiencies of 72% and 65%, respectively [113]. Addition of *M. oleifera* and PAC (10 min contact time) with coagulation (at 5 mg/L) improved removal efficiency up to 94% and 98% for PFOA and PFOS, respectively [113]. Sedimentation plus rapid sand filters achieved high-removals of PFOA (85%) and PFOS (86%) associated with particulates but low removals from the aqueous phase [115].

7.2. Oxidation Processes

Chlorine and ozone-based oxidation processes at a typical water treatment plant doses and contact times have not been effective of removing PFOA, PFOS, and other PFASs [61,116]. PFASs are resistant to chlorination or chloramination even when combined with other unit processes such as PAC and UV irradiation [61]. Advanced oxidation processes in general are ineffective for destroying PFOA and PFOS [116]. Studies of persulfate oxidation at temperatures normally encountered in the natural environment achieved an 80.5% decomposition efficiency of PFOA at 20 °C, but long reaction times are required, rendering this process unfeasible at full-scale [117].

7.3. Adsorption

Granular activated carbon (GAC) adsorption is one of the few treatment processes demonstrating significant PFAS removal from water [61,107,118]. GAC is effective in removing PFOA and PFOS in the absence of competing organics [106,119,120]. Designed appropriately, GAC will remove a contaminant to below detection limits. As the number of bed volumes of water treated increases the column effluent concentrations also increase until the contaminant level breaks through the bed. Contaminant breakthrough can be sudden or slow over time, but ultimately the influent concentration is reached at which time the column is exhausted. Once the GAC in a column has been exhausted it must be replaced and disposed of or be reactivated and reused.

When treating natural waters competitive adsorption and preloading of dissolved organic matter (DOM) must be considered. PFAS removal efficiency with GAC is highly variable with PFAS chain length based on the type of DOM and PFAS [120]. In addition, breakthrough of PFBA before PFOA and PFOS has been observed [106]. Contaminant removal is generally improved with an increase in empty bed contact time [113].

GAC is used as a biological filter to control taste and odor or to remove biologically degradable constituents such as DOM which serve as disinfection byproduct precursors. Water plants using GAC for taste and odor control or DOM removal typically replace or reactivate GAC every few years. Fresh GAC is effective at removing PFOA and PFOS. Use of GAC over one year was not effective in removing PFOA and PFOS [61,109]. PFOA and PFOS were found to increase after GAC treatment during summer months [107].

GAC filters can be costly to operate and maintain [121]. Costs are primarily determined by the flow rate of water to be treated, the influent PFOA and PFOS concentration, the presence of other PFASs to be removed, the presence of natural organic matter, GAC replacement frequency, design empty-bed contact time, and the number of bed volumes treated. Pretreatment processes may be necessary prior to GAC. To lower the cost, alternative adsorbents have been evaluated in exploratory studies. PFOS removal from water was evaluated using biochar, ash, and carbon nanotubes [122]. Carbon nanotubes exhibited higher sorption capacities for PFOS than biochar and ash.

PAC has been found effective for removing PFOA and PFOS in laboratory studies [123–125]. Compared to GAC, PAC exhibits a higher adsorption capacity and faster adsorption kinetics due to its fine particle size [123,124]. Separation and recovery of PAC from the water treated is more difficult than GAC due to its fine particle size. Coagulation, flocculation, and sedimentation are usually required to separate spent PAC from the water treated. PAC is typically used at a water plant only once, separated with other solids for further residuals treatment and disposal. In bench-studies ultrafine magnetic activated carbon consisting of Fe_3O_4 and PAC allowed separation of spend PAC from residuals using a magnet [126]. Methanol was used to regenerate the PAC for reuse up to five times.

7.4. Anion Exchange

Studies have found anion exchange to be effective for removing PFOA, PFOS, and other PFASs [107,109,110]. Anion exchange can be very effective for removing both PFASs and DOM, even with high levels of background DOM (9 mg carbon L^{-1}) [119]. When compared with GAC, anion exchange was preferred because PFAS chain length was less relevant compared to GAC, more effective at removing total DOM, and more effective at removing the most hydrophobic DOM compounds [119].

7.5. Membrane Processes

Microfiltration and ultrafiltration are low-pressure membrane processes commonly used in drinking water treatment to remove particulate matter and microorganisms. PFOA and PFOS are not removed by these processes alone due to their large membrane pore size [121].

Reverse osmosis (RO) and nanofiltration (NF) are high-pressure membrane processes not widely used for drinking water treatment but commonly used for desalination, treating brackish water, and treating wastewater for reuse [107]. RO is a proven technology for removing PFOA and PFOS, achieving up to >99% removal [116]. NF also rejects PFOA and PFOS, with about 95% rejection achieved for PFASs with molecular weights >300 g/mol [105]. NF rejection of PFOA and PFOS is highly affected by the pH of the water treated.

7.6. Treatment Process Selection

In general, a water utility may use any treatment technology acceptable to their state primacy agency to comply with a drinking water regulation. The technology selected may or may not be designated as BAT by the USEPA. However, the USEPA is required to specify BAT to meet a drinking water regulation in each proposed and final rule. A water utility seeking a variance from a regulation must agree to install a BAT.

The SDWA allows the USEPA to set an MCL or a treatment technique requirement [127]. An MCL must be set as close to the MCLG as feasible [128]. Alternatively, a treatment technique requirement obligates water systems to install the specified or equivalent treatment to the satisfaction of the state regulatory agency. The treatment technique rule is used when it is not practical or feasible to determine the level of a contaminant through laboratory testing. For example, filtration and disinfection of surface water are required to protect against waterborne disease because testing for every pathogen potentially present in surface water sources is not feasible. In the case of PFOA and PFOS, analytical methods are available as discussed above but concentration variability must also be considered. An MCL would be specified if it is feasible to reliably detect PFOA and PFOS in source and treated water during compliance monitoring. Otherwise, a treatment technique rule may be appropriate.

8. Best Available Science

Drinking water regulations established by the USEPA are to be based on the highest-quality science. Specifically, the SDWA requires [129] that:

"To the degree that an Agency action is based on science, the Administrator shall use (1) the best available, peer-reviewed science and supporting studies conducted in accordance with sound and objective scientific practices; and (2) data collected by accepted methods or best available methods (if the reliability of the method and the nature of the decision justifies use of the data".

This requirement has important implications for developing drinking water regulations for PFASs. A USEPA regulatory action may be legally challenged in the US District Court of Appeals for the District of Columba (DC) if the science upon which the regulatory action was based is arbitrary and capricious. In practice, the science behind each science-based decision will be thoroughly scrutinized by all stakeholders affected. In a prior case, the DC Circuit Court of Appeals ruled that the USEPA had failed to use the best available science in setting an MCLG of zero for chloroform. Specifically, the court ruled [130] that:

"EPA cannot reject the best available evidence simply because of the possibility of contradiction in the future by evidence unavailable at the time of the action—a possibility that will always be present."

This court decision has two important implications: (1) the agency may be acting illegally when it relies on default assumptions when the best available science supports a less (or more) conservative approached for assessing risk, and (2) the best available science is the scientific evidence available at the time of a rule-making decision. The possibility of contradiction based on further scientific data or peer review is not a legitimate basis for rejecting the science that currently exists.

The SDWA gives US federal courts jurisdiction to review USEPA actions. Though contentious, judicial review is an integral and important component of the US regulatory process. A stakeholder with standing may file a petition for judicial review of a final rule. The judicial review is limited to the administrative record and the court gives substantial deference to the USEPA when reviewing its rules. Specifically, the Court of Appeals has stated [131] that:

"We will reverse (a) USEPA action only if it is arbitrary, capricious, and abuse of discretion, or otherwise not in accordance with law (...) This highly deferential standard of review presumes Agency action to be valid (...) The rationale for deference is particularly strong when USEPA is evaluating scientific data within its technical expertise; In an area characterized by scientific and technological uncertainty, (...) this court must proceed with particular caution, avoiding all temptation to direct the Agency in its choice between rational alternatives. Despite this deferential standard, we must ensure that USEPA has examined the relevant data and has articulated an adequate explanation for its action (...) The USEPA is required to give reasonable responses to all significant comments in a rulemaking proceeding (...) We will therefore overturn a rule-making as arbitrary and capricious where USPEA has failed to respond to specific challenges that are sufficiently central to its decision".

To establish an enforceable regulation for PFOA and PFOS the detection, occurrence, exposure, health effects, human health risks, treatment technology, national costs of removal from drinking water, as well as national benefits expected from regulation must be thoroughly assessed. Decisions made in the regulatory process must be based on the best available science to avoid issuing an arbitrary and capricious regulation.

9. Costs and Benefits

At the time a new drinking water regulation is proposed the USEPA is required to publish a determination as to whether the benefits of the MCL selected justify, or do not justify, the costs of meeting the MCL based on a health risk reduction and cost analysis (HRRCA) [132]. The cost of a regulation includes capital costs for treatment installation, operation and maintenance costs, and compliance monitoring costs. The number of water systems in the United States to be affected by alternative regulatory levels is estimated. A decision tree of treatment options and unit costs is developed, and assumptions made to estimate how many water systems would install a particular process to comply with the regulation. National capital cost and operation and maintenance costs are estimated by multiplying the number of water systems using a particular treatment technology by the unit cost for that technology. The USEPA estimates costs and benefits of regulating PFOA and PFOS on a national basis using statistical projections of the number of water systems affected and unit treatment cost models. However, the cost faced by any particular water to remove PFOA and PFOS is likely to be higher than USEPA estimates.

Compliance monitoring costs are estimated based on the monitoring strategy applied to document water system compliance. Analytical methods for PFOA and PFOS require highly trained laboratory analysts and advanced laboratory equipment, and are of relatively high cost. If the standardized monitoring framework [133] is applied, initial monitoring for PFOA and PFOS would be required. Subsequent monitoring frequency would be based on initial monitoring results with waivers typically granted if concentrations are reliably and consistently below the MCL and if the water source is not vulnerable to contamination.

Adverse effects on health are quantified at the concentrations typically found in drinking water. Unquantifiable health benefits are important and considered, but quantifiable benefits form the primary bases of HRRCA. To determine the health benefits of a regulation, the difference between the prevailing ambient drinking water concentrations prior to regulation and the concentrations expected after installation of water treatment is determined on a national basis to assess exposure reduction. In the case of PFOA and PFOS, exposure reduction is expected to result in benefits regarding human serum lipids, birth weight, and serum antibodies. Reliably quantifying specific health benefits associated with exposure reduction may not be possible.

10. Regulatory Determination

The SDWA requires the USEPA to make determinations on whether or not to regulate at least five listed contaminants [134]. A notice of preliminary determination and opportunity for public comment must precede publication of final determinations. A determination to regulate a contaminant is based on findings that the criteria for health effects, occurrence, and risk reduction discussed above are met. Findings must be based on the best available public health information, including occurrence data collected during UCMR monitoring.

11. Other US Standards and Advisories

Regulatory agencies in states experiencing environmental and/or drinking water contamination of PFOA and PFOS have been very active in conducting research and setting health standards and advisories. Each state has laws governing the regulation of drinking water implemented by an agency of the state government. In general, states are not bound by SDWA legal requirements when establishing drinking water regulations but their rules must be at least as stringent as the USEPA. States typically have fewer procedural and technical constraints than the USEPA, allowing them to adopt a regulatory approach that best fits the needs of their jurisdiction.

State regulatory agencies in the United States are taking different approaches to address PFOA and PFOS in drinking water and at contaminated sites. Ten states have adopted the USEPA health advisory levels for PFOA and PFOS (Alaska, Arizona, Alabama, Colorado, Maine, Massachusetts, Michigan, New York, Rhode Island, and West Virginia) [135]. Nevada adopted a basic comparison level for PFOA and PFOS at 667 ng/L for each [136]. Exceeding a basic comparison level does not automatically designate a site as needing a response action but suggests further evaluation of health risks is warranted.

California [137] adopted non-regulatory, health-based notification levels for PFOA and PFOS of 14 ng/L and 13 ng/L, respectively. Notification levels are as set as a precautionary measure although water systems are not required to conduct monitoring. If test results exceed the notification level, then the water system must comply with state public notification requirements. When notification levels for PFOA and PFOS are exceeded and concentrations cannot be reduced below the USEPA health advisory levels removing the source from service is recommended [137].

Minnesota [6] and New Jersey [99] conducted independent risk assessments (discussed above) basing their advisory levels on different toxicological endpoints. The Minnesota health advisory levels for PFOA and PFOS are 35 ng/L and 27 ng/L, respectively [138]. The New Jersey health advisory levels for PFOA and PFOS are 14 and 13 ng/L, respectively [139].

A few states have addressed other PFASs in addition to PFOA and PFOS. New Jersey set a health advisory limit for PFNA at 13 ng/L [139]. Vermont established a health advisory limit of 20 ng/L for the sum of PFOA, PFOS, PFHxS, PFHpA, and PFNA [140].

The USEPA issued a draft interim recommendation for addressing ground water contaminated with PFOA and PFOS at sites being evaluated and addressed under federal cleanup programs [141]. For ground water contaminated with PFOA or PFOS, a screening level of 40 ng/L is proposed. A risk-based screening level is used to calculate a Hazard Quotient (HQ). Further evaluation of a contaminated site would be warranted if the PFOA or PFOS HQ is above 0.1. At sites where contaminant concentrations are below screening levels, no further action or study is generally warranted. In cases where the HQ exceeds 0.1, a decision to take remedial clean up action would be typically based on the results of a baseline risk assessment [141]. The health advisory for the combined concentration of PFOA and PFOS (70 ng/L) would be used to develop a preliminary remediation goal which may be modified as appropriate to protect human health and the environment.

The Comprehensive Environmental Response, Compensation, and Liability Act (CERCLA or Superfund) authorizes the Agency of Toxic Substances and Disease Registry (ATSDR) to prepare a toxicological profile for the hazardous substances most commonly found in facilities from the CERCLA National Priorities List and that pose the most significant potential threat to human health [67]. A draft profile for 14 PFASs was published for public comment in June 2018. Toxicological profiles are synthesis, non-regulatory documents reflecting the ATSDR's assessment of all relevant toxicologic testing and information about these substances. The ATSDR preparation of a toxicological profile is not subject to SDWA requirements.

ATSDR toxicological profile risk assessment calculations differ from the regulatory health effects and risk assessment required of the USEPA under the SDWA. If adequate data is available, ATSDR calculates a minimum risk level using a methodology similar to the USEPA calculation of the RfD. Oral minimal risk levels were proposed for PFOA (3×10^{-6} mg/kg/day), PFOS (2×10^{-6} mg/kg/day), PFHxS (2×10^{-5} mg/kg/day), and PFNA (3×10^{-6} mg/kg/day) [67]. Minimal risk levels are intended to serve as screening levels to identify hazardous waste sites where further investigation is needed. They may also be used to identify sites not expected to cause adverse health effects. Minimal risk levels are not enforceable nor intended to define clean up or action levels for ATSDR or other agencies [142].

12. International Standards

Many countries have established guidelines or action levels for PFOA and PFOS which typically are non-enforceable screening levels. Summary lists of international standards for PFASs are available [136].

A few countries have set regulatory standards. For example, Health Canada established a maximum acceptable concentration for PFOA [143] and PFOS [144] at 200 ng/L and 600 ng/L, respectively. When both are present the sum of the ratios of PFOA and PFOS detected concentrations to the corresponding maximum allowable concentration should not exceed 1. Denmark established a health-based standard for PFOA and PFOS at 100 ng/L for each [145]. Germany set a health-based drinking water standard for PFOA and PFOS at 300 ng/L for each [146]. An administrative drinking water standard for PFOA and PFOS was set at 100 ng/L for each.

13. Summary and Conclusions

PFOA and PFOS are persistent organic pollutants receiving global attention. They are two of over 4700 PFASs synthesized, many of which are in active industrial use. PFOA and PFOS have unique chemical characteristics and were manufactured and in a wide variety of industrial and consumer products used in the United States from the 1940s to 2015. Consequently, the US population has already been exposed to these substances and have low levels of them in their blood serum. Although PFOA and PFOS are no longer manufactured in the United States, the US population may still be exposed to them from legacy uses, past improper industrial waste disposal, and their transport in the environment from other countries where they are manufactured and/or used.

For drinking water regulation to be enforceable, the MCL must reflect a clear delineation between test results above and below the MCL, unaffected by variability in laboratory performance. If PFOA and PFOS are to be regulated the USEPA will establish a PQL for each. If the best treatment technology available can lower PFOA and PFOS concentrations to below the PQL, then the MCL for noncarcinogenic substances is typically set at MCLG or the PQL whichever is highest. For substances known or likely to be carcinogenic to humans the MCLG is zero, in which case the enforceable limit is set at the PQL.

PFOA and PFOS have been detected in the environment and in drinking water worldwide (Table 2). A contaminant must be known or likely to be present in drinking water at levels of public health concern to warrant a national regulation. The UCMR 3 monitoring results represent the only available statistically valid survey of national occurrence for the PFASs tested. When analyzing UCMR data, the numbers of water systems exceeding the MRL are a function of the concentration at which the MRL is established. The public health significance is evidenced by the number of water systems exceeding the HRL. The 70 ng/L HRL for PFOA was exceeded by only 13 water systems or 0.3% of the water systems tested. The 70 ng/L HRL for PFOS was exceeded by only 46 water systems or 0.9% of the water systems tested. To be regulated, a determination must be made that PFOA and PFOS are known or likely to be present in drinking water at concentrations of public health concern.

Assessing adverse health effects of PFOA and PFOS exposure involves several important scientific issues. Biomonitoring of blood serum levels indicates the body burden posed by PFOA and PFOS exposure. The RfD established for a regulation may differ from the RfD used to derive the health advisories depending on the decisions made in determining the toxicological endpoint (e.g., developmental or immunological), pharmacokinetic modeling, and interpretation of toxicological studies (e.g., LOAEL, NOAEL, uncertainty factors). Toxicological and risk assessment considerations will be a significant factor in setting MCLGs for PFOA and PFOS.

The RSC is applied to ensure an individual's total exposure from a contaminant does not exceed the RfD. The RSC is the portion of the RfD attributed to drinking water. The remainder to the RfD is allotted to other potential sources. A default value of 20% was used to derive the health advisories based on exposure to the general population. Unless additional data become available, this same RSC default value will be used to develop a national regulation for PFOA and PFOS.

The SDWA requires the MCL to be set as close to the MCLG as is feasible. "Feasible" means feasible with the use of the best technology, treatment techniques, and other means which the USEPA finds after examination for efficacy under field conditions and not solely under laboratory conditions (taking cost into consideration) [102]. Few technologies meet this feasibility criterion. GAC adsorption, anion exchange, and reverse osmosis remove PFOA and PFOS. The effectiveness of treatment technology and the PQL usually have the most influence on the MCL determination. If PFOA and PFOS are regulated, the effectiveness and cost of treatment technology or other means being considered must be determined. The USEPA is required to make a determination whether the national benefits justify the national cost for each final national primary drinking water regulation.

The best available peer-reviewed science and supporting studies "conducted in accordance with sound and objective scientific practices" are to be used when a regulatory decision is based on science. Only data collected by accepted methods or best available methods are to be used if the reliability of the method and the nature of the decision justifies use of the data [129].

At the time a new drinking water regulation is proposed, the USEPA is required to publish a determination as to whether the benefits of the MCL selected justify, or do not justify, the costs of meeting the MCL based on a health risk reduction and cost analysis (HRRCA) [132]. The cost of a regulation includes capital costs for treatment installation, operation and maintenance costs, and compliance monitoring costs.

The quantifiable and unquantifiable benefits of the regulation must be estimated if a regulation is developed. Adverse effects on health are quantified at the concentrations before and after treatment. Quantified benefits form the primary basis of HRRCA, which presents the costs and benefits of each alternative MCL under consideration.

A decision whether to regulate PFOA and PFOS is anticipated as part of the Regulatory Determination 4. The USEPA decided to not regulate 24 contaminants in prior Regulatory Determinations 1, 2, and 3. [147–149] A specific rationale for the decision to not regulate was provided for each contaminant. Prior decisions provide a backdrop and precedent for the decision whether to regulate PFOA and PFOS. For example, aldrin and dieldrin are not regulated because their use had been banned and they had "a low frequency and low level of occurrence in drinking water." The percentage of water systems exceeding the HRL for aldrin and dieldrin were 0.494% and 0.2%, respectively. Hexachloropentadiene occurs in public water systems but is not regulated because it did not occur at a frequency or level of public health concern. Manganese is not regulated because it was "generally not considered to be very toxic when ingested with the diet and drinking water accounts for a relatively small proportion of manganese intake".

The frequency of occurrence and potential health risks are a significant factor for the PFOA and PFOS determination. In Regulatory Determination 1 metribuzin and naphthalene were not regulated because they were "not known to occur (...) at a level of public health concern" and were infrequently detected. Other contaminants for which a determination was made to not regulate include the dacthal mono- and di-acid degradates; 1,1-dichloro-2,2-bis(p-chlorophenyl)ethylene (DDE); 1,3-dichloropropene; 2,4-dinitrotoluene; 2,6-dinitrotoluene; s-ethyl dipropylthiocarbamate (EPTC); fonofos; terbacil; 1,1,2,2-tetrachloroethane; dimethoate, 1,3-dinitrobenzene, terbufos; and terbufos sulfone. [147–149].

A majority of states have accepted and are applying the USEPA health advisory limits for PFOA and PFOS. Several states performed independent risk assessments and set state-specific advisory limits, especially states where significant environmental contamination had occurred due to improper industrial waste disposal. International agencies, ATSDR, and state agencies independently develop advisory limits within the constraints of their prevailing legal requirements. The USEPA PFOA and PFOS health advisory limits and any future drinking water regulations are developed under the SDWA [15], APA [16], and the agency's regulatory policies, which are more stringent than the requirements faced by other agencies.

PFOA and PFOS contamination will continue to be addressed by state agencies and USEPA regional offices regardless of whether these contaminants are regulated nationally. Contamination first discovered occurred prior to enactment of present environmental laws and were addressed through litigation [36,37]. Several thousand lawsuits were filed against the manufacturers of PFOA and PFOS resulting in large monetary settlements [150–152]. Affected water systems and consumers can access a wealth of information on PFOA and PFOS from the USEPA and many state agencies.

Several agencies have been actively addressing PFOA and PFOS contamination for well over a decade. Much has been learned but extensive research is still needed to answer key questions. Two congressionally mandated research grants have been initiated. Investigators at Oregon State University and North Caroline State University have been granted US$2.6 million to define and predict the toxicity of PFASs, in work that is due for completion in April 2022 [153]. A research team from the Colorado School of Mines, Duke University, Michigan State University, North Carolina State University, and the University of Colorado at Denver has been granted US$2.45 million to develop actionable data on the fate, transport, bioaccumulation, and exposure of a large suite of PFASs in nationally representative PFAS-impacted communities, in a study due for completion in April 2022 [154]. As knowledge gaps are filled by these and other studies, better decisions can be made on how best to reduce total PFAS exposure.

Author Contributions: This article was written and produced entirely by the author from May to August 2019 while the author was a visiting scholar at the Chung Yuan Christian University, Taoyuan City, Taiwan, The author examined over 320 articles, reports and government documents citing the most relevant in this review. Tables were prepared from data extracted from the original literature cited. Figure 1 is in the public domain.

Funding: This research received no external funding.

Conflicts of Interest: The author declares no conflict of interest.

References

1. KEMI The Swedish Chemicals Agency. Occurrence and Use of Highly Fluorinated Substances and Alternatives. Report 7/15. 2015. Available online: https://www.kemi.se/global/rapporter/2015/report-7-15-occurrence-and-use-of-highly-fluorinated-substances-and-alternatives.pdf (accessed on 22 September 2019).
2. Wang, Z.; Cousins, I.T.; Scheringer, M.; Buck, R.C.; Hungerbuhler, K. Global emission inventories for C_4–C_{14} perfluoroalkyl carboxylic acid (PFCA) homologues from 1951 to 2030, Part I: Production and emissions from quantifiable sources. *Environ. Int.* **2014**, *70*, 62–75. [CrossRef] [PubMed]
3. Wang, Z.; Cousins, I.T.; Scheringer, M.; Buck, R.C.; Hungerbuhler, K. Global emission inventories for C_4–C_{14} perfluoroalkyl carboxylic acid (PFCA) homologues from 1951 to 2030, part II: The remaining pieces of the puzzle. *Environ. Int.* **2014**, *69*, 166–176. [CrossRef] [PubMed]
4. Organization for Economic Co-Operation and Development (OECD). Toward a New Comprehensive Global Database of Per- and Polyfluoroalkyl Substances (PFASs): Summary Report on Updating the OECD 2007 List of Per- and Polyfluoroalkyl Substances (PFASs). Series on Risk Management No. 39, ENV/JM/MONO(2018)7. 4 May 2018. Available online: https://www.oecd.org/chemicalsafety/portal-perfluorinated-chemicals/ (accessed on 19 July 2019).
5. Herrick, R.L.; Buckholz, J.; Biro, F.M.; Calafat, A.M.; Ye, X.; Xie, C.; Pinney, S.M. Polyfluoroalkyl substance exposure in the Mid-Ohio River Valley, 1991–2012. *Environ. Pollut.* **2017**, *228*, 50–60. [CrossRef] [PubMed]
6. Patzke, J. Investigating Drinking Water Contamination in Ohio by Per- and Polyfluoroalkyl Substances. Ohio EPA, Division of Drinking and Ground Waters: Columbus, OH, USA, 22 October 2018.
7. Goeden, H.M.; Greene, C.W.; Jacobus, J.A. A transgenerational toxicokinetic model and its use in derivation of Minnesota PFOA water guidance. *J. Expo. Sci. Environ. Epidemol.* **2019**, *29*, 183–195. [CrossRef] [PubMed]
8. USEPA. Drinking Water Contaminant Candidate List 3 (CCL3)—Final. *Fed. Regist.* **2009**, *74*, 51850–51862.
9. USEPA. Drinking Water Health Advisory for Perfluorooctanoic Acid (PFOA). EPA 822-R-16-005; Office of Water: Washington, DC, USA, May 2016. Available online: https://www.epa.gov/sites/production/files/2016-05/documents/pfoa_health_advisory_final_508.pdf (accessed on 22 September 2019).
10. USEPA. Drinking Water Health Advisory for Perfluorooctane Sulfonate (PFOS). EPA 822-R-16-004; Office of Water: Washington, DC, USA, May 2016. Available online: https://www.epa.gov/sites/production/files/2016-05/documents/pfos_health_advisory_final_508.pdf (accessed on 22 September 2019).
11. Cordner, A.; De La Rosa, V.Y.; Schaider, L.A.; Rudel, R.A.; Richter, L.; Brown, P. Guideline levels for PFOA and PFOS in drinking water: The role of scientific uncertainty, risk assessment decisions, and social factors. *J. Expo. Sci. Environ. Epidemol.* **2019**, *29*, 157–171. [CrossRef] [PubMed]

12. Reed, J.; Stabenow, D.; Warren, E.; Durban, R.; Manchin, J.; Harris, K.; Gillibrand, K.; Murray, P.; Carper, T.; Coons, C.; et al. Letter from United States Senators to EPA Administrator Scott Pruitt. 13 April 2018. Available online: https://drive.google.com/file/d/1LgpWUVI-wfvSW90LtTzjymSNm_BAZTj1/view (accessed on 22 September 2019).
13. USEPA. EPA's Per- and Polyfluoroalkyl Substances (PFAS) Action Plan. EPA 823R18004; 2009. Available online: https://www.epa.gov/sites/production/files/2019-02/documents/pfas_action_plan_021319_508compliant_1.pdf (accessed on 22 September 2019).
14. USEPA. Fact Sheet: EPA's PFAS Action Plan: A Summary of Key Actions. 2019. Available online: https://www.epa.gov/sites/production/files/2019-02/documents/pfas_action_factsheet_021319_final_508compliant.pdf (accessed on 22 September 2019).
15. Public Law 104–182, The Safe Drinking Water Act Amendments of 1996. Available online: https://www.congress.gov/bill/104th-congress/senate-bill/1316 (accessed on 22 September 2019).
16. Administrative Conference. *A Guide to Federal Agency Rulemaking*, 2nd ed.; Office of the Chairman, Administrative Conference of the United States: Washington, DC, USA, 1991.
17. USEPA Science Advisory Board. *Reducing Risk: Setting Priorities and Strategies for Environmental Protection*; SAB-E-90-021; USEPA Science Advisory Board: Washington, DC, USA, 1990.
18. Public Law 99–339. 1986 Safe Drinking Water Act Amendments. Sec. 1412(b)(3)(A). Available online: https://www.congress.gov/104/plaws/publ182/PLAW-104publ182.pdf (accessed on 22 September 2019).
19. Public Law 104–182. The Safe Drinking Water Act Amendments of 1996. Sec.1412(b)(1)(D). Available online: https://www.congress.gov/104/plaws/publ182/PLAW-104publ182.pdf (accessed on 22 September 2019).
20. Buck, R.C.; Franklin, J.; Berger, U.; Conder, J.M.; Cousins, I.T.; De Voogt, P.; Jensen, A.A.; Kannan, K.; Mabury, A.S.; Van Leeuwen, S.P. Perfluoroalkyl and polyfluoroalkyl substances in the environment: Terminology, classification, and origins. *Integr. Environ. Assess. Manag.* **2011**, *7*, 513–541. [CrossRef]
21. Lindstrom, A.B.; Strynar, M.J.; Libelo, E. Polyfluorintated compounds: Past, present, and future. *Environ. Sci. Technol.* **2011**, *45*, 7954–7961. [CrossRef]
22. Barzen-Hanson, K.A.; Roberts, S.C.; Choyke, S.; Oetjen, K.; McAlees, A.; Riddell, N.; McCrindle, R.; Ferguson, P.L.; Higgins, C.P.; Field, J.A. Discovery of 40 Classes of Per- and Polyfluoroalkyl Substances in Historical Aqueous Film-Forming Foams (AFFFs) and AFFF-Impacted Groundwater. *Environ. Sci. Technol.* **2017**, *51*, 2047–2057. [CrossRef]
23. Jian, J.-M.; Guo, Y.; Zeng, L.; Liang-Ying, L.; Lu, X.; Wang, F.; Zeng, E.Y. Global distribution of perfluorochemicals (PFCs) in potential human exposure source—A review. *Environ. Int.* **2017**, *108*, 51–62. [CrossRef]
24. Kunacheva, C.; Shivakoti, B.R.; Lien, N.P.H.; Harada, H. Worldwide surveys of perfluorooctane sulfonate (PFOS) and perfluorooctanoic acid (PFOA) in water environment in recent years. *Water Sci. Technol.* **2012**. [CrossRef] [PubMed]
25. Centers for Disease Control and Prevention. *Fourth National Report on Human Exposure to Environmental Chemicals, Updated Tables, January 2019*; Dept. of Health and Human Services, Centers for Disease Control and Prevention (CDC): Atlanta, GA, USA, 2019. Available online: https://www.cdc.gov/exposurereport/ (accessed on 15 April 2019).
26. Agency for Toxic Substances and Disease Registry. *Perfluoroalkyl and Polyfluoroalkyl Substances (PFAS) in the U.S. Population*; Dept. of Health and Human Services: Atlanta, GA, USA, 2017.
27. Interstate Technology Regulatory Council (ITRC). *History and Use of Per- and Polyfluoroalkyl Substances (PFAS)*; ITRC: Washington, DC, USA, 2017.
28. Wang, Z.W.; DeWitt, J.C.; Higgins, C.P.; Cousins, I.T. A never-ending story of per- and polyfluoroalkyl substances (PFASs)? *Environ. Sci. Technol.* **2017**, *51*, 2508–2518. [CrossRef] [PubMed]
29. Nakayama, S.F.; Yoshikane, M.; Onoda, Y.; Nishihama, Y.; Iwai-Shimada, M.; Takagi, M.; Kobayashi, Y.; Isobe, T. Worldwide trends in tracing poly- and perfluoroalkyl substances (PFAS) in the environment. *Trends Anal. Chem.* **2019**. [CrossRef]
30. Shoemaker, J.A.; Grimmett, P.E.; Boutin, B.K. *Method 537, Determination of Selected Perfluoro Alkyl Acids in Drinking Water by Solid Phase Extraction and Liquid Chromatography/Tandem Mass Spectrometry (LC/MS/MS)*; Version 1.1, EPA/600/R-08/092; USEPA Office of Research and Development: Cincinnati, OH, USA, 2009.

31. Shoemaker, J.A.; Tettenhorst, D.R. *Method 537.1, Determination of Selected Per- and Polyfluorinated Alkyl Subtances in Drinking Water by Solid Phase Extraction and Liquid Chromatography/Tandem Mass Spectrometry (LC/MS/MS)*; Version 1.0, EPA/600/R-18/352; USEPA Office of Research and Development: Cincinnati, OH, USA, 2018.
32. USEPA. Drinking Water Contaminant Candidate List 4—Final. *Fed. Regist.* **2016**, *81*, 81099–81114.
33. USEPA. Revisions to the Unregulated Contaminant Monitoring Rule (UCMR 3) for Public Water Systems. *Fed. Regist.* **2012**, *77*, 26072–26101.
34. USEPA. *The Third Unregulated Contaminant Monitoring Rule (UCMR 3): Data Summary*; EPA 815-S-17-001; Office of Water: Washington, DC, USA, 2017.
35. New Jersey Department of Environmental Protection. Interim Practical Quantitation Level (PQL) Determination to Support Interim Specific Ground Water Quality Standard Development for Perfluorooctanoic Acid (PFOA). Division of Science and Research; 6 March 2019. Available online: https://www.nj.gov/dep/dsr/supportdocs/PFOA_PQL.pdf (accessed on 19 June 2019).
36. Rich, N. The Lawyer Who Became DuPont's Worst Nightmare. *The New York Times Magazine*, 6 January 2016. Available online: https://www.nytimes.com/2016/01/10/magazine/the-lawyer-who-became-duponts-worst-nightmare.html(accessed on 19 June 2019).
37. Mordock, J. Taking on duPont: Illnesses, Deaths Blamed on Pollution from W. VA Plant. *Delaware Online*. 1 April 2016. Available online: https://www.delawareonline.com/story/news/2016/04/01/dupont-illnesses-deaths-c8/81151346/ (accessed on 19 June 2019).
38. Minnesota Pollution Control Agency. Perfluorochemicals (PFCs). Undated Webpage. Available online: https://www.pca.state.mn.us/waste/perfluorochemicals-pfcs (accessed on 19 June 2019).
39. Hu, X.C.; Andrews, D.Q.; Lindstrom, A.B.; Bruton, T.A.; Schaider, L.A.; Granjean, P.; Lohmann, R.; Carignan, C.C.; Blum, A.; Balan, S.A.; et al. Detection of Poly- and Perfluoroalkyl Substances (PFASs) in U.S. Drinking Water Linked to Industrial Sites, Military Fire Training Areas, and Wastewater Treatment Plants. *Environ. Sci. Technol. Lett.* **2016**, *3*, 344–350. [CrossRef]
40. Liu, C.; Gin, K.Y.H.; Chang, V.W.C.; Goh, B.P.L. Novel perspectives on the bioaccumulation of PFCs–The concentration dependency. *Environ. Sci. Technol.* **2011**, *45*, 9758–9764. [CrossRef]
41. Zhu, H.; Kannan, K. Distribution and partitioning of perfluoroalkyl carboxylic acids in surface soil, plants, and earthworms at a contaminated site. *Sci. Total Environ.* **2019**, 647–961. [CrossRef]
42. Thompson, J.; Eaglesham, G.; Mueller, J. Concentrations of PFOS, PFOA and other perfluorinated alkyl acids in Australian drinking water. *Chemosphere* **2011**, *83*, 1320–1325. [CrossRef]
43. Szabo, D.; Coggan, T.L.; Robson, T.C.; Currell, M.; Clarke, B.O. Investigating recycled water use as a diffuse source of per- and polyfluoroalkyl substances (PFASs) to groundwater in Melbourne, Australia. *Sci. Total Envrion.* **2018**, *644*, 1409–1417. [CrossRef]
44. Thompson, J.; Roach, A.; Eaglesham, G.; Bartkow, M.E.; Edge, K.; Mueller, J.F. Perfluorinated alkyl acids in water, sediment and wildlife from Sydney Harbor and surroundings. *Mar. Pollut. Bull.* **2011**, *62*, 2869–2875. [CrossRef] [PubMed]
45. Furdui, V.I.; Crozier, P.W.; Reiner, E.J.; Mabury, S.A. Trace level determination of perfluorinated compounds in water by direct injection. *Chemosphere* **2008**, *73*, 524–530. [CrossRef] [PubMed]
46. Mak, Y.L.; Taniyasu, S.; Yeung, L.W.Y.; Lu, G.; Jin, L.; Yang, Y.; Lam, P.K.S.; Kannan, K.; Yamashita, N. Perfluorinated Compounds in Tap Water from China and Several Other Countries. *Environ. Sci. Technol.* **2009**, *43*, 4824–4829. [CrossRef] [PubMed]
47. Kaboré, H.A.; Duy, S.V.; Munoz, G.; Méité, L.; Desrosiers, M.; Liu, J.; Sory, T.K.; Sauvé, S. Worldwide drinking water occurrence and levels of newly-identified perfluoroalkyl and polyfluoroalkyl substances. *Sci. Total Environ.* **2018**, *616–617*, 1089–1100. [CrossRef] [PubMed]
48. Chen, C.; Lu, Y.; Zhang, X.; Geng, J.; Wang, T.; Shi, Y.; Hu, W.; Li, J. A review of spatial and temporal assessment of PFOS and PFOA contamination in China. *Chem. Ecol.* **2009**, *25*, 163–177. [CrossRef]
49. So, M.K.; Miyake, Y.; Yeung, W.Y.; Ho, Y.M.; Taniyasu, S.; Rostowski, P.; Yamashita, N.; Zhou, B.S.; Shi, X.J.; Wang, J.X.; et al. Perfluorinated compounds in the Pearl River and Yangtze River of China. *Chemosphere* **2007**, *68*, 2085–2095. [CrossRef] [PubMed]
50. Chen, S.; Zhou, Y.; Meng, J.; Wang, T. Seasonal and annual variations in removal efficiency of perfluoro alkyl substances by different wastewater treatment processes. *Environ. Pollut.* **2018**, *242*, 2059–2067. [CrossRef] [PubMed]

51. Zhang, W.; Zhang, Y.; Taniyasu, S.; Yeung, L.W.Y.; Lam, P.K.S.; Wang, J.; Li, X.; Yamashita, N.; Dai, J. Detection and fate of perfluoroalkyl substances in municipal wastewater treatment plants in economically developed areas of China. *Environ. Pollut.* **2013**, *176*, 10–17. [CrossRef] [PubMed]
52. Li, Y.; Li, J.; Zhang, L.; Huang, Z.; Liu, Y.; Wu, N.; He, J.; Zhang, Z.; Zhang, Y.; Niu, Z. Perfluoroalkyl acids in drinking water of China 2017: Distribution characteristics, influencing factors and potential risks. *Environ. Int.* **2019**, *123*, 87–95. [CrossRef]
53. Jin, Y.H.; Liu, W.; Sato, I.; Nakayama, S.F.; Sasaki, K.; Saito, N.; Tsuda, S. PFOS and PFOA in environmental and tap water in China. *Chemosphere* **2009**, *77*, 605–611. [CrossRef]
54. Ahrens, L.; Felizeter, S.; Strum, R.; Xie, Z.; Ebinghaus, R. Polyfluorinated compounds in waste water treatment plant effluents and surface waters along the River Elbe, Germany. *Mar. Pollut. Bull.* **2009**, *58*, 1326–1333. [CrossRef] [PubMed]
55. Sunantha, G.; Vasudenvan, N. Assessment of perfluorooctanoic acid and perfluorooctane sulfonate in surface water. *Mar. Pollut. Bull.* **2016**, *109*, 612–618. [CrossRef] [PubMed]
56. Takagi, F.; Adachi, F.; Miyano, K.; Tanaka, H.; Mimura, M.; Watanabe, I.; Tanabe, S.; Kannan, K. Perfluorooctanesulfonate and perfluorooctanoate in raw and treated tap water from Osaka, Japan. *Chemosphere* **2008**, *72*, 1409–1412. [CrossRef] [PubMed]
57. Zafeiraki, E.; Costopoulou, D.; Vassiliadou, I.; Leondiadis, L.; Dassenakis, E.; Traag, W.; Hoogenboom, R.L.A.P.; van Leeuwn, S.P.J. Determination of perfluoroalkylated substances (PFASs) in drinking water from the Netherlands and Greece. *Food Addit. Contam. A* **2015**, *32*, 2048–2057. [CrossRef] [PubMed]
58. Brandsma, S.H.; Koekkoek, J.C.; van Velzen, M.J.M.; de Boer, J. The PFOA substitute GenX detected in the environment near a fluoropolymer manufacturing plant in the Netherlands. *Chemosphere* **2019**, *220*, 493–500. [CrossRef] [PubMed]
59. Lin, A.Y.-C.; Panchangam, S.C.; Lo, C.-C. The impact of semiconductor, electronics and optoelectronic industries on downstream perfluorinated chemical contamination in Taiwan rivers. *Environ. Pollut.* **2009**, *157*, 1365–1372. [CrossRef] [PubMed]
60. Boone, J.S.; Vigo, C.; Boone, T.; Byrne, C.; Ferrario, J.; Benson, R.; Donohue, J.; Simmons, J.E.; Kolpin, D.W.; Furlong, E.T.; et al. Per- and polyfluoroalkyl substances in source and treated drinking water of the United State. *Sci. Total Environ.* **2019**, *653*, 359–369. [CrossRef]
61. Quiñones, O.; Snyder, S.A. Occurrence of Perfluoroalkyl Carboxylates and Sulfonates in Drinking Water and Related Waters from the United States. *Environ. Sci. Technol.* **2009**, *43*, 9089–9095. [CrossRef]
62. Hartz, M. PFAS Monitoring in a Post Health Advisory World—What Should We Be Doing? Presented at AWWA New York Section Conference. 2017. Available online: https://nysawwa.org/docs/presentations/2017/FINAL-PFAS%20Monitoring%20in%20Post%20health%20Advisory%20World-What%20Should%20We%20Be%20Doing-2017.pdf (accessed on 22 September 2019).
63. USEPA. *Provisional Health Advisory for Perfluorooctanoic Acid (PFOA) and Perfluorooctane Sulfonate (PFOS)*; Office of Water: Washington, DC, USA, 8 January 2009. Available online: https://www.epa.gov/sites/production/files/2015-09/documents/pfoa-pfos-provisional.pdf (accessed on 19 June 2019).
64. USEPA. *Framework for Human Health Risk Assessment to Inform Decision Making*; EPA/100/R-14/001; Risk Assessment Forum: Washington, DC, USA, April 2014. Available online: https://www.epa.gov/sites/production/files/2014-12/documents/hhra-framework-final-2014.pdf (accessed on 22 September 2019).
65. USEPA. *Health Effects Support Document for Perfluorooctanoic Acid (PFOA)*; EPA 822-R-16-003; Office of Water: Washington, DC, USA, 2016.
66. USEPA. *Health Effects Support Document for Perfluorooctane Sulfonate (PFOS)*; EPA 822-R-16-002; Office of Water: Washington, DC, USA, 2016.
67. ATSDR. *Toxicological Profile for Perfluoroalkyls–Draft for Public Comment*; US Dept of Health and Human Services: Atlanta, GA, USA, 2018. Available online: https://www.atsdr.cdc.gov/toxprofiles/tp.asp?id=1117&tid=237 (accessed on 22 September 2019).
68. Dong, Z.; Bahar, M.M.; Jit, J.; Kennedy, B.; Priestly, B.; Ng, J.; Lamb, D.; Liu, Y.; Duan, L.; Naidu, R. Issues raised by the reference doses for perfluorooctane sulfonate and perfluorooactanoic acid. *Environ. Int.* **2017**, *105*, 86–94. [CrossRef]
69. Post, G.B.; Gleason, J.A.; Cooper, K.R. Key scientific issues in developing drinking water guidelines for perfluoroalkyl acids: Contaminants of emerging concern. *PLoS Biol.* **2017**, *15*, e2002855. [CrossRef]

70. Olesn, G.W.; Burris, J.M.; Ehresham, D.J.; Froehlich, J.W.; Seacat, A.M.; Butenhoff, J.L.; Zobel, L.R. Half-life of Serum Elimination of Perfluorooctanesulfonate, Perfluorohexanesulfonate, and Perfluorooctanoate in Retired Fluorochemical Production Workers. *Envrion. Health Perspect.* **2007**, *115*, 1298–1305. [CrossRef]
71. Li, Y.; Fletcher, T.; Mucs, D.; Scott, K.; Lindh, C.H.; Tallving, P.; Jakobsson, K. Half-lives of PFOS, PFHxS and PFOA after end of exposure to contaminated drinking water. *Occup. Environ. Med.* **2018**, *75*, 46–51. [CrossRef]
72. Loccisano, A.E.; Campbell, J.L.; Andersen, M.E.; Clewell, H.J. Evaluation and prediction of pharmacokinetics of PFOA and PFOS in the monkey and human using a PBPK model. *Regul. Toxicol. Pharm.* **2011**, *59*, 157–175. [CrossRef]
73. Zhang, S.; Kang, Q.; Peng, H.; Ding, M.; Zhao, F.; Zhou, Y.; Dong, Z.; Zhang, H.; Yang, M.; Tao, S.; et al. Relationship between perfluoroactanoate and perfluorooctane sulfonate blood concentrations in the general population and routine drinking water exposure. *Environ. Int.* **2019**, *126*, 54–60. [CrossRef]
74. Olsen, G.W.; Burris, J.M.; Berlew, M.M.; Mandel, J.H. Epidemiologic assessment of worker serum perfluoroactanesulfonate (PFOS) and perfluorooctanoate (PFOA) concentrations and medical surveillance examinations. *J. Occup. Environ. Med.* **2003**, *45*, 260–270. [CrossRef]
75. Sakr, C.J.; Lenonard, R.C.; Kreckmann, K.H.; Slade, M.D.; Cullen, M.R. Longitudinal study of serum lipids and liver enzymes in workers with occupational exposure to ammonium perfluorooctanoate. *J. Occup. Environ. Med.* **2007**, *49*, 872–879. [CrossRef]
76. Emmett, E.A.; Zhang, H.; Shofer, F.S.; Freeman, D.; Rodway, N.V.; Desai, C.; Shaw, L.M. Community exposure to perfluorooctanoate: Relationships between serum levels and certain health parameters. *J. Occup. Environ. Med.* **2006**, *48*, 771–779. [CrossRef]
77. Steenland, K.; Jin, C.; MacNeil, J.; Lally, C.; Ducatman, A.; Vieira, V.; Fletcher, T. Predictors of PFOA levels in a community surrounding a chemical plant. *Environ. Health Perspect.* **2009**, *117*, 1083–1088. [CrossRef]
78. Worley, R.R.; Moore, S.M.; Tierney, B.C.; Ye, X.; Calafate, A.M.; Campbell, S.; Woudneh, M.B.; Fisher, J. Per- and polyfluoroalkyl substances in human serum and urine samples from a residentially exposed community. *Environ. Int.* **2017**, *106*, 135–143. [CrossRef]
79. Daly, E.R.; Chan, B.P.; Talbot, E.A.; Nassif, J.; Bean, C.; Cavallo, S.J.; Metcalf, E.; Simone, K.; Woolf, A.D. Per- and polyfluoroalkyl substances (PFAS) exposure assessment in a community exposed to contaminated drinking water, New Hampshire, 2015. *Int. J. Hyg. Envir. Heal.* **2018**, *221*, 569–577. [CrossRef]
80. Olsen, G.W.; Mair, D.C.; Lange, C.C.; Harrington, L.M.; Church, T.R.; Goldberg, C.L.; Herron, R.M.; Hanna, H.; Nobiletti, J.B.; Rios, J.A.; et al. Per- and polyfluoroalkyl substances (PFAS) in American Red Cross adult blood donors, 2000–2015. *Environ. Res.* **2017**, *157*, 87–95. [CrossRef]
81. USEPA. *A Review of the Reference Dose and Reference Concentration Processes*; EPA/630/P-02/0002F; Risk Assessment Forum: Washington, DC, USA, 2002. Available online: https://www.epa.gov/sites/production/files/2014-12/documents/rfd-final.pdf (accessed on 22 September 2019).
82. Lau, C.; Thibodeaux, J.R.; Hanson, R.G.; Narotsky, M.G.; Rogers, J.M.; Lindstrom, A.B.; Strynar, M.J. Effects of perfluorooctanoic acid exposure during pregnancy in the mouse. *Toxicol. Sci.* **2006**, *90*, 510–518. [CrossRef]
83. Luebker, D.J.; Case, R.G.; York, R.G.; Moore, J.A.; Hansen, K.J.; Butenoff, J.L. Two-generation reproduction and cross-foster studies of perfluorooctanesulfonate (PFOS) in rats. *Toxicology* **2005**, *215*, 126–148. [CrossRef]
84. Nicole, W. PFOA and Cancer in a Highly Exposed Community. *Environ. Health Perspect.* **2013**, *121*, A340. [CrossRef]
85. USEPA. *Guidelines for Carcinogen Risk Assessment*; EPA/630/P-03/001F; Risk Assessment Forum: Washington, DC, USA, March 2005. Available online: https://www.epa.gov/sites/production/files/2013-09/documents/cancer_guidelines_final_3-25-05.pdf (accessed on 22 September 2019).
86. Butenhoff, J.L.; Kennedy, G.L.; Chang, S.C.; Olsen, G.W. Chronic dietary toxicity and carcinogenicity study with ammonium perfluorooctanotae in Sprague-Dawley rats. *Toxicology* **2012**, *298*, 1–13. [CrossRef]
87. Arrieta-Cortes, R.; Farias, P.; Hoyo-Vadillo, C.; Kleiche-Dray, M. Carcinogenic risk of emerging persistent organic pollutant perfluorooctane sulfonate (PFOS): A proposal of classification. *Regul. Toxicol. Pharm.* **2017**, *83*, 66–80. [CrossRef]
88. Begley, T.H.; White, K.; Honigfort, P.; Twaroski, M.L.; Neches, R.; Walker, R.A. Perfluorochemicals: Potential sources of and migration from food packaging. *Food Addit. Contam.* **2005**, *22*, 1023–1031. [CrossRef]
89. USFDA. Indirect Food Additives: Paper and Paperboard Components. *Fed. Regist.* **2016**, *81*, 5–8.

90. Fraser, A.J.; Webster, T.F.; Watkins, D.J.; Strynar, M.J.; Kato, K.; Calafat, A.M.; Vieira, V.M.; McClean, M.D. Polyfluorinated compounds in dust from homes, offices, and vehicles as predictors of concentrations in office workers' serum. *Environ. Int.* **2013**, *60*, 128–136. [CrossRef]
91. Trudel, D.; Horowitz, L.; Wormuth, M.; Scheringer, M.; Cousins, I.T.; Hungerbühler, K. Estimating Consumer Exposure to PFOS and PFOA. *Risk Anal.* **2008**, *28*, 251–269. [CrossRef]
92. Kim, D.-H.; Lee, J.-H.; Oh, J.-E. Assessment of individual-based perfluoroalkyl substances exposure by multiple human exposure sources. *J. Hazard. Mater.* **2019**, *365*, 26–33. [CrossRef]
93. Wu, N.; Cai, D.; Guo, M.; Li, M.; Li, X. Per- and polyfluorinated compounds in sales women's urine linked to indoor dust in clothing shops. *Sci. Total Environ.* **2019**, *667*, 594–600. [CrossRef]
94. Franko, J.; Meade, B.J.; Frasch, H.F.; Barbero, A.M.; Anderson, S.E. Dermal Penetration Potential of Perfluorooctanoic Acid (PFOA) in Human and Mouse Skin. *J. Toxicol. Environ. Health A* **2012**, *75*. [CrossRef]
95. Gebbink, V.G.; Berger, U.; Cousins, I.T. Estimating human exposure to PFOS isomers and PFCA homologues: The relative importance of direct and indirect (precursor) exposure. *Environ. Int.* **2015**, *74*, 160–169. [CrossRef]
96. National Toxicology Program. *NTP Monograph on Immunotoxicity Associated with Exposure to Perfluorooctanoic Acid (PFOA) or Perfluorooctane Sulfonate (PFOS)*; Office of Health Assessment and Translation: Research Triangle Park/Durham, NC, USA, 2016.
97. Peden-Adams, M.M.; Keller, J.M.; Eudaly, J.G.; Berger, J.; Gilkeson, G.S.; Keil, D.E. Suppression of humnoral immunity is mice following exposure to perfluorooctane sulfonate. *Toxicol. Sci.* **2008**, *104*, 144–154. [CrossRef]
98. Lilienthal, H.; Dieter, H.H.; Holzer, J.; Wilhelm, M. Recent experimental results of effects of perfluoroalkyl substances in laboratory animals–Relation to current regulations and guidance values. *Int. J. Hyg. Environ. Health* **2017**, *220*, 766–775. [CrossRef]
99. Pachkowski, B.; Post, G.B.; Stern, A.H. The derivation of a Reference Dose (RfD) for Perfluorooctane sulfonate (PFOS) based on immune suppression. *Environ. Res.* **2019**, *171*, 452–469. [CrossRef]
100. Dong, G.H.; Zhang, Y.H.; Zheng, L.; Liu, W.; Jin, Y.H.; He, Q.C. Chronic effects of perfluorooctanesulfonate exposure on immunotoxicity in adult male C57B1/6 mice. *Arch. Toxicol.* **2009**, *83*, 805–815. [CrossRef]
101. Chang, E.T.; Adami, H.-O.; Boffetta, P.; Wedner, H.J.; Mandel, J.S. A critical review of perfluorooactanoate and perfluorooctanesulfonate exposure and immunological health conditions in humans. *Crit. Rev. Toxicol.* **2016**, *46*, 279–331. [CrossRef]
102. 1986 SDWA Amendments Sec. 1412(b)(4)(D). Available online: https://www.congress.gov/bill/99th-congress/senate-bill/124 (accessed on 22 September 2019).
103. Rayne, S.; Forest, K. Perfluoroalkyl sulfonic and carboxylic acids: A critical review of physicochemical properties, levels and patterns in waters and wastewaters, and treatment methods. *J. Environ. Sci. Health A* **2009**, *44*, 1145–1199. [CrossRef]
104. Vecitis, C.D.; Park, H.; Cheng, J.; Mader, B.T.; Hoffman, M.R. Treatment technologies for aqueous perfluoroocatanesulfonate (PFOS) and perfluorooctanoate (PFOA). *Front. Environ. Sci. Eng. China* **2009**, *3*, 129–151. [CrossRef]
105. Eschauzier, C.; Beerendonk, E.; Scholte-Veenendall, P.; Voogt, P.D. Impact of Treatment Processes on the Removal of Perfluoroalkyl Acids from Drinking Water Production Chain. *Environ. Sci. Technol.* **2012**, *46*, 1708–1715. [CrossRef]
106. Appleman, T.D.; Higgins, C.P.; Quinones, O.; Vanderford, B.J.; Kolstad, C.; Zeigler-Holady, J.C.; Dickenson, E.R.V. Treatment of poly- and perfluoroalkyl substances in U.S. full-scale water treatment systems. *Water Res.* **2014**, *51*, 246–255. [CrossRef]
107. Rahman, M.F.; Peldszus, S.; Anderson, W.B. Behaviours and fate of perfluoroalkyl and polyfluoroalkyl substances (PFASs) in drinking water treatment: A review. *Water Res.* **2014**, *50*, 318–340. [CrossRef]
108. Kucharzyk, K.H.; Darlington, R.; Benotti, M.; Deeb, R.; Hawley, E. Novel treatment technologies for PFAS compounds: A critical review. *J. Environ. Manag.* **2017**, *204*, 757–764. [CrossRef]
109. Takagi, S.; Adachi, F.; Miyano, K.; Koizumi, Y.; Tanaka, H.; Watanabe, I.; Tanabe, S.; Kannan, K. Fate of perfluoroocatanesulfonate and perfluorooctanoate in drinking water treatment processes. *Water Res.* **2011**, *45*, 3925–3932. [CrossRef]
110. Thompson, J.; Eaglesham, G.; Reungoat, J.; Poussade, Y.; Bartkow, M.; Lawrence, M.; Mueller, J.F. Removal of PFOS, PFOA and other perfluoroalkyl acids at water reclamation plants in South East Queensland Australia. *Chemosphere* **2011**, *82*, 9–17. [CrossRef]

111. Xiao, F.; Simcik, J.F.; Gulliver, J.S. Mechanisms for removal of perfluorooctane sulfonate (PFOS) and perfluorooctanoate (PFOA) from drinking water by conventional and enhanced coagulation. *Water Res.* **2013**, *47*, 49–56. [CrossRef]
112. Bao, Y.; Niu, J.; Xu, Z.; Gao, D.; Shi, J.; Sun, X.; Huang, Q. Removal of perfluorooctanoate sulfonate (PFOS) and perfluorooctanoate (PFOA) form water by coagulation: Mechanisms and influencing factors. *J. Colloid Interface Sci.* **2014**, *434*, 59–64. [CrossRef]
113. Pramanik, B.K.; Pramanki, S.K.; Suja, F. A comparative study of coagulation, granular- and powdered activated carbon for the removal of perfluorooctane sulfonate and perfluorooctanoate in drinking water treatment. *Environ. Technol.* **2015**, *36*, 2610–2617. [CrossRef]
114. Deng, S.; Zhou, Q.; Yu, G.; Huan, J.; Fan, Q. Removal of perfluorooctanoate from surface water by polyaluminum chloride coagulation. *Water Res.* **2011**, *45*, 1774–1780. [CrossRef]
115. Kunacheva, C.; Fuji, S.; Tanaka, S.; Boontanon, S.K.; Poothong, S.; Wongwatthana, T.; Shivakoti, B.R. Perfluorinated compounds contamination in tap water and bottled water in Bangkok, Thailand. *J. Water Supply Res. Technol.* **2010**, *59*, 345–354. [CrossRef]
116. Espana, V.A.A.; Mallavarapu, M.; Naidu, R. Treatment Technologies for aqueous perfluoroocatanesulfonate (PFOS) and perfluorooctanoate (PFOA): A critical review with an emphasis on field testing. *Environ. Technol. Innov.* **2015**, *4*, 168–181. [CrossRef]
117. Lee, Y.-C.; Lo, S.-L.; Lin, Y.-L. Persulfate oxidation of perfluorooctanoic acid under the temperatures of 20–40 °C. *Chem. Eng. J.* **2012**, *198–199*, 27–32. [CrossRef]
118. McNamara, J.D.; Franco, R.; Mimna, R.; Zappa, L. Comparison of Activated Carbons for Removal of Perfluorinated Compounds from Drinking Water. *J. AWWA* **2018**, *110*. [CrossRef]
119. McCleaf, P.; Englund, S.; Ostlund, A.; Lindegren, K.; Wiberg, K. Removal efficiency of multiple poly- and perfluoroalkyl substances (PFASs) in drinking water using granular activated carbon (GAC) and anion exchange)AE) column tests. *Water Res.* **2017**, *120*, 77–87. [CrossRef]
120. Kothawala, K.H.; Kohler, S.J.; Ostund, A.; Wiberg, K.; Ahrens, L. Influence of dissolved organic matter concentration and composition of the removal efficiency of perfluoroalkyl substances (PFASs) during drinking water treatment. *Water Res.* **2017**, *121*, 320–328. [CrossRef]
121. Hoslett, J.; Massara, T.M.; Malamis, S.; Ahmad, D.; van den Boogaert, I.; Katsou, E.; Ahmad, B.; Ghazal, H.; Simons, S.; Wrobel, L.; et al. Surface water filtration using granular media and membranes: A review. *Sci. Total Environ.* **2008**, *639*, 1268–1282. [CrossRef]
122. Chen, X.; Xia, X.; Wang, X.; Qiao, J.; Chen, H. A comparative study on sorption of perfluoroactane sulfonate. *Chemosphere* **2011**, *83*, 1313–1319. [CrossRef]
123. Qu, Y.; Zhang, C.; Li, F.; Bo, X.; Liu, G.; Zhou, Q. Equilibrium and kinetics study on the adsorption of perfluorooctanoic acid from aqueous solution onto powdered activated carbon. *J. Hazard. Mater.* **2009**, *169*, 146–152. [CrossRef]
124. Yu, Q.; Zhang, R.; Deng, S.; Huang, J.; Yu, G. Sorption of perfluorooctane sulfonate and perfluorooctanoate on activated carbons and resins: Kinetic and isotherm study. *Water Res.* **2009**, *43*, 1150–1158. [CrossRef] [PubMed]
125. Hansen, M.C.; Børresen, M.H.; Schlabach, M.; Cornelissen, G. Sorption of perfluorinated compounds from contaminated water to activated carbon. *J. Soils Sediments* **2010**, *10*, 179. [CrossRef]
126. Meng, P.; Fang, X.; Maimaiti, A.; Yu, G.; Deng, S. Efficient removal of perfluroinated compounds from water using a regenerable magnetic activated carbon. *Chemosphere* **2019**, *224*, 187–194. [CrossRef] [PubMed]
127. Public Law 104–182, The Safe Drinking Water Act Amendments of 1996, Section 1412(b)(7). Available online: https://www.congress.gov/104/plaws/publ182/PLAW-104publ182.pdf (accessed on 22 September 2019).
128. Public Law 104–182, The Safe Drinking Water Act Amendments of 1996, Section 1412(b)(4)(B). Available online: https://www.congress.gov/104/plaws/publ182/PLAW-104publ182.pdf (accessed on 22 September 2019).
129. Public Law 104–182, The Safe Drinking Water Act Amendments of 1996, Section 1412(b)(4)(D). Available online: https://www.congress.gov/104/plaws/publ182/PLAW-104publ182.pdf (accessed on 22 September 2019).
130. U.S. Court of Appeals. *Chlorine Chemistry Council and Chemical Manufacturers Association v. EPA.*; U.S. Court of Appeals, Case 99–1627; District of Columbia Circui: Washington, DC, USA, 2000.

131. U.S. Court of Appeals. *International Fabricare Institute for Itself and on Behalf of its Members v. USEPA.*; U.S. Court of Appeals, Case 91–1838; District of Columbia Circuit: Washington, DC, USA, 1994.
132. Public Law 104–182, The Safe Drinking Water Act Amendments of 1996, Section 1412(b)(3)(C). Available online: https://www.congress.gov/104/plaws/publ182/PLAW-104publ182.pdf (accessed on 22 September 2019).
133. USEPA. *Standardized Monitoring Framework, Office of Water*; EPA 570/F-91-045; USEPA: Washington, DC, USA, 1991.
134. Public Law 104–182, The Safe Drinking Water Act Amendments of 1996, Section 1412(b)(1)(B)(ii). Available online: https://www.congress.gov/104/plaws/publ182/PLAW-104publ182.pdf (accessed on 22 September 2019).
135. Orange County Water District. *PFOA and PFOS Occurrence in the Santa Anna River Watershed*; OCWD Presentation to SAWPA EC Task Force; Orange County Water District: Fountain Valley, CA, USA, 2019.
136. ITRC. PFAS Fact Sheet Table 4–1. Standards and Guidance Values for PFAS in Groundwater, Drinking Water, and Surface Water/Effluent (Wastewater). June 2019. Available online: https://pfas-1.itrcweb.org/fact-sheets/ (accessed on 22 September 2019).
137. California State Water Resources Control Board. Perfluorooctanoic Acid (PFOA) and Perfluorooctanesulfonic Acid (PFOS) Webpage, Division of Drinking Water. 2019. Available online: https://www.waterboards.ca.gov/drinking_water/certlic/drinkingwater/PFOA_PFOS.html (accessed on 22 September 2019).
138. Minnesota Department of Health. Human Health-Based Water Guidance Table. 2019. Available online: https://www.health.state.mn.us/communities/environment/risk/guidance/gw/table.html (accessed on 13 August 2109).
139. New Jersey Department of Environmental Protection. Site Remediation Program Website, Contaminants of Emerging Concern. 13 March 2019. Available online: https://www.nj.gov/dep/srp/emerging-contaminants/ (accessed on 13 August 2109).
140. Vermont Department of Health. *Memo from Schwer, C. to Chapmann, M. and Englander, D. Perfluorooctanoic acid (PFOA) and Perfluorooctanesulfonic acid (PFOS)*; Vermont Drinking Water Health Advisory: Burlington, VT, USA, 2016.
141. USEPA. Draft Interim Recommendations to Address Groundwater Contaminated with Perfluorooctanoic Acid and Perfluoroctane Sulfonate. April 2019. Available online: https://www.epa.gov/sites/production/files/2019-04/documents/draft_interim_recommendations_for_addressing_groundwater_contaminated_with_pfoa_and_pfos_public_comment_draft_4-24-19.508post.pdf (accessed on 22 September 2019).
142. ATSDR. Toxic Substances Portal, Minimal Risk Levels (MRLs)—For Professionals. 21 June 2018. Available online: https://www.atsdr.cdc.gov/mrls/index.asp (accessed on 31 July 2019).
143. Health Canada. Guidelines for Canadian Drinking Water Quality: Guideline Technical Document. Perfluoroactane Acid (PFOA). December 2018. Available online: https://www.canada.ca/en/health-canada/services/publications/healthy-living/guidelines-canadian-drinking-water-quality-technical-document-perfluorooctanoic-acid/document.html (accessed on 31 July 2019).
144. Health Canada. Guidelines for Canadian Drinking Water Quality: Guideline Technical Document. Perfluorooctane Sulfonate (PFOS). December 2018. Available online: https://www.canada.ca/en/health-canada/services/publications/healthy-living/guidelines-canadian-drinking-water-quality-guideline-technical-document-perfluorooctane-sulfonate/document.html#1.0 (accessed on 31 July 2019).
145. Larsen, P.B.; Giovalle, E. *Danish Ministry of the Environment. Perfluoroalkylated Substances: PFOA, PFOS and PFOSA: Evaluation of Health Hazards and Proposal of a Health Based Quality Criterion for Drinking Water, Soil and Ground Water*; Environmental project No. 1665; The Danish Environmental Protection Agency: Copenhagen, Denmark, 2015; Available online: http://www2.mst.dk/Udgiv/publications/2015/04/978-87-93283-01-5.pdf (accessed on 23 May 2019).
146. German Ministry of Health. Assessment of PFOA in the Drinking Water of the German Hochsauerlandkreis. Provisional Evaluation of PFT in Drinking Water with the Guide Substances Perfluorooctanoic acid (PFOA) and Perfluorooctane Sulfonate (PFOS) as Examples. 2006. Available online: http://www.umweltbundesamt.de/sites/default/files/medien/pdfs/pft-in-drinking-water.pdf (accessed on 23 May 2019).
147. USEPA. *Contaminant Candidate List Regulatory Determination Support Document for Aldrin and Dieldrin*; EPA-815-R-03-010; Office of Water: Washington, DC, USA, 2003.

148. USEPA. *Regulatory Determinations Support Document for Selected Contaminants from the Second Drinking Water Contaminant Candidate List (CCL2)*; EPA 815-R-08-012; Office of Water: Washington, DC, USA, 2008.
149. USEPA. Announcement of Final Regulatory Determinations for Contaminants on the Third Drinking Water Contaminant Candidate List. *Fed. Regist.* **2016**, *81*, 13–19.
150. Mordock, J. DuPont, Chemours to Pay $670 Million over PFOA Suits. *Delaware News Journal, Delaware Online*. 17 February 2017. Available online: http://delonline.us/2kCechH (accessed on 13 August 2019).
151. Minnesota Pollution Control Agency. 3M and PFCs: 2018 Settlement Website, Undated. Available online: https://www.pca.state.mn.us/waste/3m-and-pfcs-2018-settlement (accessed on 13 August 2019).
152. Reisch, M.S. 3M settles PFAS suit in Alabama. *Chem. Eng. News* **2019**, *97*, 12.
153. USEPA. System Toxicological Approaches to Define and Predict the Toxicity of Per- and Poly-Fluoroalkyl Substances. EPA Grant No. R839481. Available online: https://cfpub.epa.gov/ncer_abstracts/index.cfm/fuseaction/display.abstractDetail/abstract/10950/report/0 (accessed on 22 September 2019).
154. USEPA. PFAS United: Poly- and Perfluoroalkyl Substances- US National Investigation of Transport and Exposure from Drinking Water and Diet. EPA Grant No. R839482; 1 May 2019. Available online: https://cfpub.epa.gov/ncer_abstracts/index.cfm/fuseaction/display.abstractDetail/abstract/10951/report/0 (accessed on 22 September 2019).

Article

PPCP Monitoring in Drinking Water Supply Systems: The Example of Kárený Waterworks in Central Bohemia

Zbyněk Hrkal [1,2,*], Pavel Eckhardt [1], Anna Hrabánková [1], Eva Novotná [1] and David Rozman [1,2,*]

1. T.G. Masaryk Water Research Institute, p.r.i, 160 00 Praha 6-Dejvice, Czech Republic; eckhardt@vuv.cz (P.E.); hrabankova@vuv.cz (A.H.); novotna@vuv.cz (E.N.); rozman@vuv.cz (D.R.)
2. Institute of Hydrogeology, Engineering Geology and Applied Geophysics, Faculty of Sciences, Charles University, Albertov 6, 12 843 Prague 2, Czech Republic

* Correspondence: hrkal@vuv.cz; Tel.: +420-606-079-144

Received: 15 November 2018; Accepted: 8 December 2018; Published: 13 December 2018

Abstract: The Kárany waterworks supplies drinking water to about one-third of Prague, the capital city of the Czech Republic with a population of more than 1 million. The combination of two technologies—bank infiltration and artificial recharge—are used for production of drinking water. The two-year monitoring of PPCPs (pharmaceuticals and personal care products) at monthly intervals observed temporal changes in 81 substances in the source river and groundwater, and the efficacy of contamination removal depended on the treatment technology used. The results showed a very wide range of PPCPs discharged from the waste water treatment plant at Mladá Boleslav into the Jizera River at concentrations ranging from ng/L to μg/L. Acesulfame and oxypurinol in concentrations exceeding 100 ng/L systematically occurred, and then a few tens of ng/L of carbamazepine, sulfamethoxazole, primidone, and lamotrigine were regularly detected at the water outlet using the artificial recharge for production of drinking water. Bank infiltration was found more efficient in removing PPCP substances at the Kárany locality where none of the monitored substances was systematically detected in the mixed sample.

Keywords: emerging pollutants; wastewater; drinking water; bank infiltration; artificial recharge

1. Introduction

Only a few years ago, most experts in water management had only very vague ideas about the occurrence and amount of the so-called micropollutants, that are substances contained in water at extremely low concentrations in the order of nanograms per liter. Also, for this reason until now, these substances are not dealt with in the Czech or European legislation for drinking water. However, very fast development of sophisticated analytical laboratory methods disclosed a number of pharmaceuticals in water. The abbreviation PPCP (pharmaceuticals and personal care products) for this very diverse group of substances is now used [1–4].

PPCPs include, for example, pharmaceuticals that enter waste waters from sewers. Current purification technologies are mostly inefficient, and in the extreme case, some substances are not removed at all during the water purification process. During the monitoring Czech-Norwegian Research Programme project AQUARIUS (Assessing water quality improvement options concerning nutrient and pharmaceutical contaminants in rural watersheds) undertaken at the pilot site of Horní Beřkovice in Central Bohemia, PPCPs such as hydrochlorothiazide, sulfamethoxazole, sulphapyridine, sulphanilamide, carbamazepine, including its metabolite carbamazepine-10,11-epoxide, were systematically detected downgradient from the mechanical and biological waste water treatment plant between the years 2015 and 2016 [5,6]. These substances were also recorded at very low concentrations in the order of tens to hundreds of nanograms per liter (only carbamazepine, gabapentin, and hydrochlorothiazide were detected at concentrations in the order of micrograms per liter). Similar results were observed in constructed wetlands in the catchment of the Želivka water reservoir [7–9].

Similar studies carried out in the USA [10], Great Britain [11], Germany [12], and Switzerland [13] showed identical problems with the low efficiency of traditional wastewater treatment plants. The technological solution of eliminating the majority of forms of PPCPs from wastewater exists and comprises the use of activated carbon. The high efficacy of this technology was demonstrated in a number of experiments by Rodrigues et al. [14] and Rivera-Utrilla et al. [15] but it is a relatively high cost method.

Considering the extremely low detected concentrations, the negative impact of PPCPs on human health is still speculative [16,17]. No long-term clinical studies have shown any negative effects of PPCPs contained in drinking water on the human organism. For this reason, we are still working with the term potential or unquantified risk [18,19].

Since 2013, the European Commission has implemented PPCPs to legislation by establishing a watch list of substances for EU-wide monitoring in the field of water policy pursuant to Directive 2008/105/EC of the European Parliament and of the Council. The document was amended in 2015 (EU 2015/495) [20] and currently includes anti-inflammatory pharmaceutical diclofenac, hormones 17-beta-estradiol (E2), 17-alpha-ethinylestradiol (EE2), estrone (E1), antibiotics erythromycin, clarithromycin, azithromycin, and several other substances like selected insecticides and herbicides.

In the Czech Republic, PPCPs in drinking water have not been adequately addressed. For this reason, the present study is focused particularly on the detection of PPCPs in the process of drinking water treatment in the Káraný waterworks. The monitoring system is designed to clarify the behavior of micropollutants on their way from the source to the waterworks.

2. Materials and Methods

2.1. Characteristics of the Káraný Pilot Site

The above-mentioned findings led to the initiation of the detailed monitoring of the quality of drinking water at the Káraný waterworks supplying the capital city of Prague. The selected pilot area is a unique locality. Along a 32 km long stretch of the Jizera River from Mladá Boleslav to the Sojovice weir, as shown in Figure 1, the behavior of pharmaceuticals was studied. Finally, the efficacy of removing PPCPs during bank infiltration and artificial recharge was assessed.

The waterworks at Káraný operates on the principle of combining two independent drinking water treatment technologies. The first one is now historic, but still a perfectly functioning project of bank infiltration built between 1906 and 1913. It consists of 685 wells of a depth ranging from 8 to 12 m, spaced 20 to 40 m apart, situated in the sand-gravel fluvial terraces ca. 250 m from the bank of the Jizera River, as shown in Figure 1. The total capacity of this system is up to 1000 L/s.

Another section of the waterworks started in 1968 and relies on artificial recharge [21] during which the surface water from the Jizera River is, after a simple mechanical treatment, pumped into infiltration ponds, as shown in Figure 2, from where it moves into about 20 m thick sandy fluvial sediments. The water table is at an average level of 10 to 14 m below ground so that there is, in the unsaturated zone, sufficient storage space for seepage water. At a distance of approximately 200 m from the infiltration ponds, there is a system of large-diameter wells with a total capacity of up to 900 L/s. The tapped water is a mixture of infiltrated water and original groundwater in a sandy-gravel terrace inflowing from the east towards the Jizera River. Water balance model studies [22] assume that 20% to 30% of groundwater participates in the resulting mixture, while the remaining 70% to 80% consists of water from the artificial recharge. However, these proportions may vary depending on the operating conditions of the waterworks.

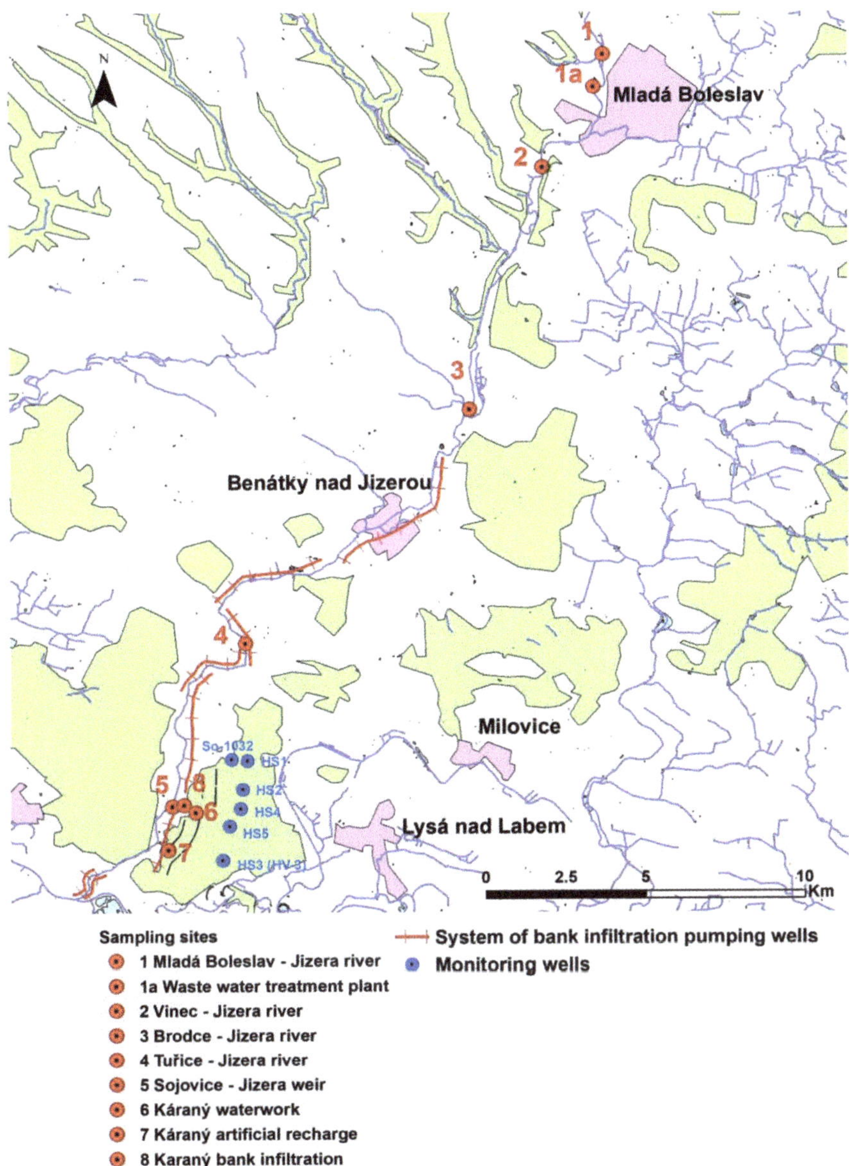

Figure 1. Sampling sites on the Jizera River.

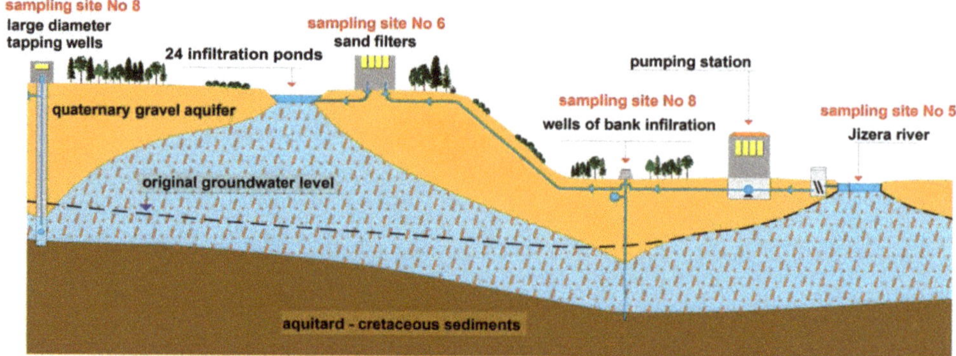

Figure 2. Scheme of Káraný waterworks (modified from Skalický [22]).

A total of 94 PPCPs or their metabolites were monitored on a monthly basis during two consecutive years. During July and August 2017, when the water discharge of the Jizera River was expected to decrease, and therefore the concentrations of the monitored substances increased, the frequency of sampling was shortened to once a week. In addition, changes in PPCP concentrations in river water over a period of 24 h were monitored by sampling at 2-h intervals one time in August 2017.

Sampling was carried out at nine sites, as shown in Figure 1. Profiles No. 1–No. 5 represent sampling sites on the Jizera River. Profile No. 1 upstream of the Mladá Boleslav town characterizes the quality of water flowing from the upper reaches of the river above the municipal wastewater treatment plant. Analyses from profile No. 1a define the quality of the purified urban wastewater (including a psychiatric hospital) discharged in the Jizera River. The results from profiles No. 2, No. 3, No. 4, and No. 5 show changes in water quality during its flow in the watercourse. Profile No. 5 is a key point for the Káraný waterworks, because it is a sampling site for the subsequent artificial recharge process.

The sixth sampling site represents the water quality after its mechanical treatment and prior to infiltration into the Quaternary aquifer. The analysis at monitoring site No. 7 characterizes the water mixture from all currently operating wells downgradient from the artificial recharge. It actually defines the qualitative changes that take place in the process of artificial recharge. However, the analyses do not always take into account the mixing from all infiltration ponds. For operational reasons, there may occur a situation when some of the wells can be temporarily shut down.

Analyses from the eighth sampling site represent a mixture of water from bank infiltration wells. However, these are not all used permanently, depending on the input parameters of the tapped water, and the desired yield, so that some parts of the water system are cut off.

To clarify the quality of groundwater in the Quaternary aquifer inflowing from the east, a group of water monitoring boreholes marked 9, 10, 11, and 12, as shown in Figure 1, were also sampled in October 2018.

2.2. Analytical Methods

The analyses of the collected wastewater, surface and groundwater, mud and soil samples were carried out according to a validated procedure in the Vltava catchment laboratory.

Samples were taken in a 60 mL amber glass vial (filling only half of the volume). Samples were stored in a freezer (in an inclined position). On the day of analysis, samples were defrosted at a maximum of 30 °C. It was necessary to continue the analysis procedure immediately after defrosting.

Two methods were developed for analysis of PPCPs—Method A (ESI+ (Electrospray ionization) mode)) and Method B (ESI− mode). The samples were centrifuged in headspace vials for 5 min at about 3500 rpm. Then, 1.50 grams of the samples were weighed in a 2 mL vial on the analytical balance. Then, 1.5 µL of formic acid (Method A) or 1.5 µL of acetic acid (Method B) was added into the sample. An isotope dilution was performed in the next step. Deuterized internal standards of d10-carbamazepine, d6-sulfamethoxazole, d3-iopromide, and 13C2-erythomycin (Method A), or d3-ibuprofen, d4-diclofenac, d3-naproxen, d5-chloramphenicol, and d3-iopamidol (Method B) were used.

PPCPs were separated and detected by LC-MS/MS (Liquide Chromatography-Mass Spectrometry/ Mass Spectrometry) methods based on direct injection of the sample into a chromatograph. A 1200 Ultra High-Performance Liquid Chromatograph (UHPLC, Agilent Technologies, Santa Clara, CA, USA) tandem with 6410 Triple Quad Mass Spectrophotometer (MS/MS, Agilent Technologies, Santa Clara, CA, USA) from Agilent Technologies were used in ESI+ or ESI− mode.

Method A (ESI+)—the separation was carried out on a Zorbax Eclipse XDB-C18 analytical column (100 × 4.6 mm, 3.5 µm particle size, Agilent Technologies, Santa Clara, CA, USA). The mobile phase consisted of methanol and water with 0.1% formic acid and 5 mM ammonium formate as the mobile phase additives. The flow rate was 0.25 mL min^{-1}. Injection volume was 0.50 mL.

Method B (ESI−)The separation was carried out on a Zorbax Eclipse XDB-C18 analytical column (150 × 4.6 mm, 3.5 µm particle size). The mobile phase consisted of methanol and water with 0.05% acetic acid as the mobile phase additive. The flow rate was 0.25 mL min^{-1}. Injection volume was 1 mL.

The samples for non-steroidal anti-inflammatory drugs were first acidified by acetic acid, filtered through a 0.45 µm cellulose filter, and mixed with an internal standard solution. The internal standard solution was prepared of deuterized solids (98% purity) and water from a UHQ (Ultra High Quality) system. The SPE (Solid Phase Extraction) was performed on a high-performance liquid chromatography column.

The samples for the estrogen group analysis were first acidified by hydrochloric acid to pH 2. Then, acetonitrile solutions of particular substances were added as internal standards, the pH was raised to 7.8, and samples were filtered through a 1 µm glass-fiber filter. The SPE was performed on a conditioned solid phase extraction disc. The analytes were eluted by acetonitrile, dried, dissolved in hexane and dichloromethane, cleaned on a florisil column, and transferred to the solution for LC/MS (Liquide Chromatography-Mass Spectrometry/ Mass Spectrometry) detection.

3. Results

3.1. Source Area of PPCPs in the River Water

All smaller cities in the upper reaches of the Jizera River are a source of PPCPs contained in the river water. However, absolutely crucial is the city of Mladá Boleslav with a population of more than 44,000; especially the psychiatric hospital at Kosmonosy that is connected to the local sewage treatment plant. Its importance is documented in Figure 3, which demonstrates the varied composition of PPCPs that are at concentrations exceeding thousands of nanograms per liter and being discharged into the Jizera River. Only substances that appeared in analyses over the reference period of two years and being above the detection limit in more than 25% of cases were included in the survey. Therefore, accidental or sporadic occurrences were excluded from the statistics.

The results clearly show the impact of the Mladá Boleslav wastewater treatment plant on the quality of water in the Jizera River. Oxypurinol and telmisartan occur systematically in purified waste water at concentrations of tens of micrograms per liter, and the other four drugs, namely diclofenac, tramadol, lamotrigine, and hydrochlorothiazide, enter the stream at concentrations at the micrograms per liter level. As for analyses of the other 44 pharmaceuticals, their concentrations are systematically ranging from a single nanogram to hundreds of nanograms per liter.

Outflow from the wastewater treatment plant usually is in tens of L/s, while the Jizera River discharge at Mladá Boleslav is in m^3 per second range for most of the year. Nevertheless, as follows from the comparison of long-term concentrations of micropollutants on the profile above the treatment plant (profile No. 1), and at the sampling site below it (profile No. 2), there is a distinct impact on the quality of surface water. Figure 4 shows the qualitative changes that occur in the long term (2-year average) on a relatively short stretch of 8 km of the Jizera River between profiles No. 1 and No. 2. Of the 44 substances systematically detected in the river water below the wastewater treatment plant on profile No. 2, 18 substances showed elevated content, while only three of them decreased slightly, and concentrations of the remaining substances were found more or less stable. The impact of discharging treated wastewater in the case of oxypurinol, telmisartan, and iomeprol is quite conclusive. These substances were either absent on profile No. 1 (iomeprol) or only detected at low concentrations (oxypurinol, telmisartan).

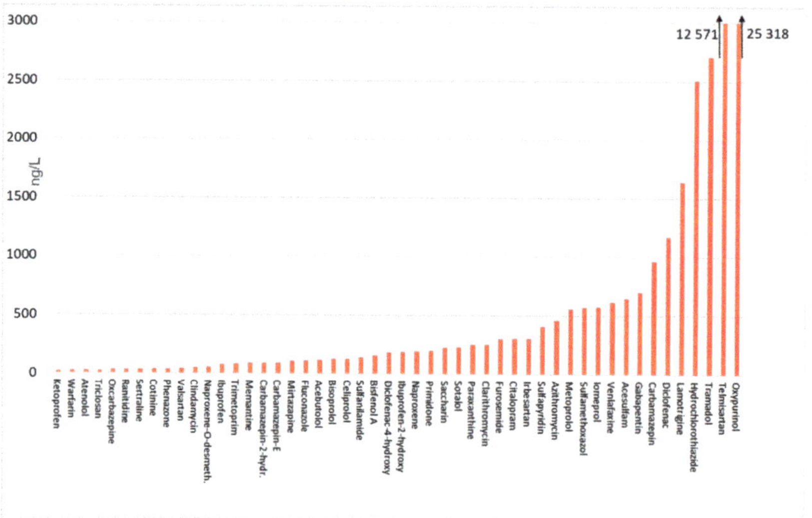

Figure 3. Average content of pharmaceuticals and personal care products (PPCPs) that are systematically discharged from the wastewater treatment plant in Mladá Boleslav.

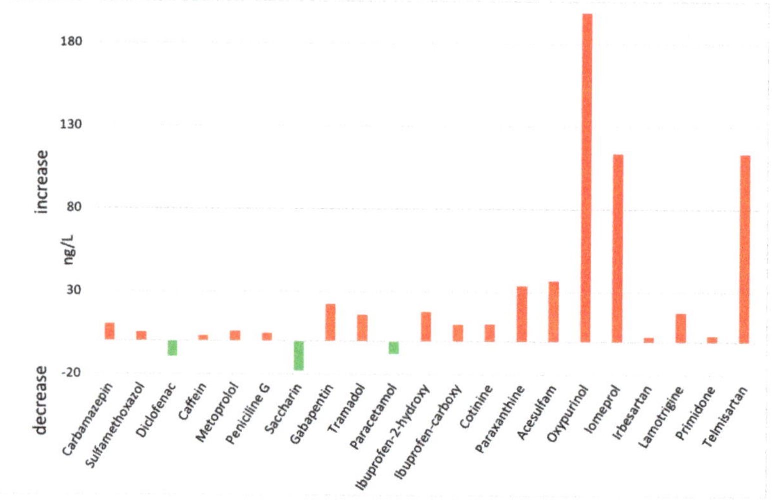

Figure 4. Changes in long-term concentrations of monitored PPCPs on the profiles above and below the wastewater treatment plant at Mladá Boleslav (in red are substances whose concentrations have increased in this section of the watercourse, the green ones have fallen).

3.2. Changes in PPCP Concentrations in River Water between Mladá Boleslav and the Weir in Sojovice

The monitored profiles on the Jizera River in the section between Mladá Boleslav and the weir in Sojovice—where the water is collected for artificial recharge—document the processes resulting in natural attenuation of pharmaceuticals in the river water. Only small tributaries with very limited water discharge join this stretch of the Jizera River, so they do not significantly affect the water balance of the Jizera River. The municipal wastewater treatment plant of the city of Benátky nad Jizerou, with a population of 7400, appears to be the only new source of pharmaceuticals. This fact is very clear in comparison to the two-year monitoring of average river water quality on profiles No. 2 Vinec and No. 5 Sojovice weir. While the concentrations of most substances were found to be decreasing downstream of Mladá Boleslav in a stretch about 32 km long, as shown in Figure 5, their content between the profiles No. 1 and No. 2 were observed to be increasing in general.

The section of the Jizera River around the Sojovice weir plays a key role for part of the Káraný waterworks, which uses artificial recharge for the production of drinking water. This water is the source for further treatment, more or less using natural purification processes. Despite the predominant decreasing trend of most pollutants in the Jizera River downstream of Mladá Boleslav, their detected number and concentrations in absolute values remain at relatively high levels (Figure 6). The waterworks in Káraný infiltrates a very varied mixture of 26 micropollutants whose average concentration in four cases exceeds 200 ng/L.

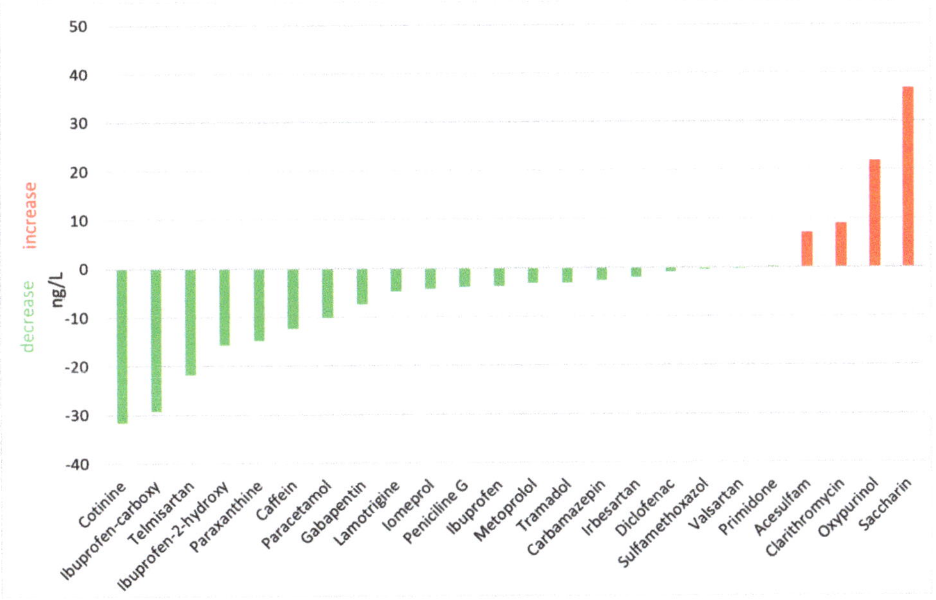

Figure 5. Changes in long-term concentrations of monitored PPCPs between profiles No. 2 and No. 5 Sojovice weir.

Interesting results were obtained along profile No. 5 during 8 August 2017. The data acquired in two-hour consecutive intervals showed a large variability in concentrations of some groups of micropollutants over a short period of one single day, as shown in Figure 7. The monitored substances can be divided into several groups. The first group consisted of carbamazepine, sulfamethoxazole, metoprolol, gabapentin, or tramadol; the concentrations of which did not change over the last 24 h, and whose contents are subject only to longer-term changes.

Conversely, ibuprofen, ibuprofen-2-hydroxy, or diclofenac concentrations in surface water changed even during a few hours. Ibuprofen showed a relatively obvious trend with gradual onset from night-time concentration values of about 40 ng/L, which doubled shortly after midday, and was followed by a slight decline. Ibuprofen-2-hydroxy had a maximum at 140 ng/L at 06:00, followed by a systematic drop until night-time. Diclofenac fluctuated during the day without any clear trend.

There was a very specific behavior of saccharin and caffeine. Both micropollutants reflect the habits and lifestyle of the common population, first appearing in surface water as late as at around 10:00, and the last caffeine was detected in surface water at 20:00. Saccharin was detected two hours later. Thus, it is evident that the residence time of these two substances in water was very short, and their occurrence was closely linked to the local source.

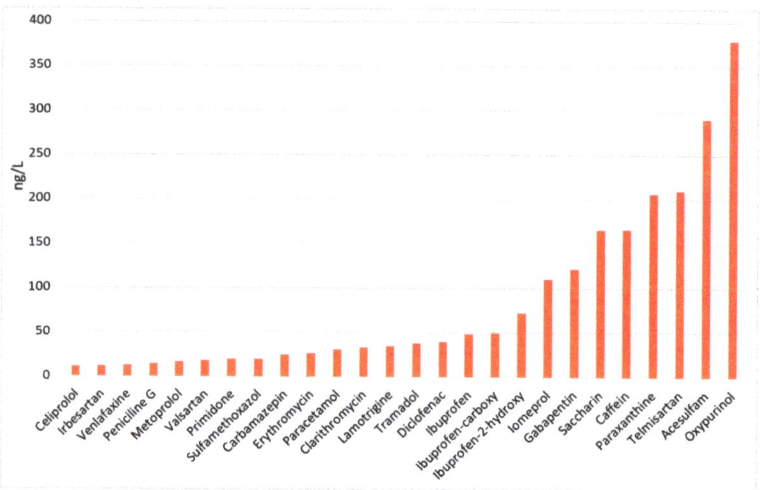

Figure 6. Long-term average concentrations of PPCPs on profile No. 5 Sojovice weir.

The data suggest that to achieve a reliable comparability of input data for subsequent statistical processing, it is necessary to adhere to a uniform sampling time.

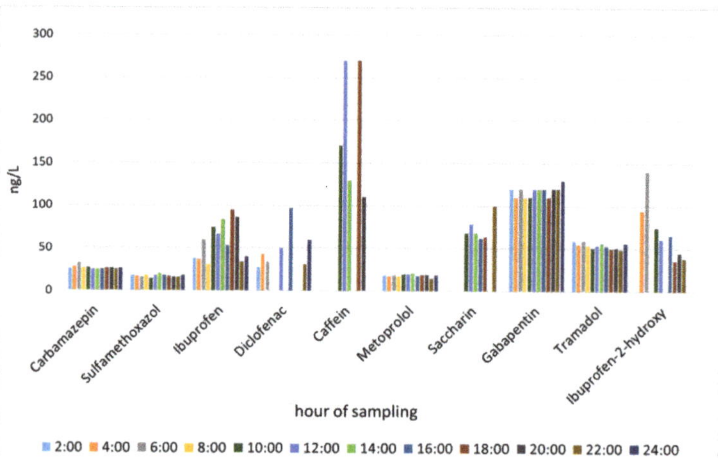

Figure 7. Time-related changes in concentrations of selected PPCPs on profile No. 5 during August 2017.

The quality of source water for artificial recharge and bank infiltration is subject to relatively considerable time-related changes in water discharge. Figure 8 clearly demonstrates the dilution ability of the selected three pharmaceuticals that systematically occur in the stream on profile No. 5. The content of metoprolol, carbamazepine, and tramadol at elevated water discharges (10 m^3/s) were always below the detection limit of the analytical method used.

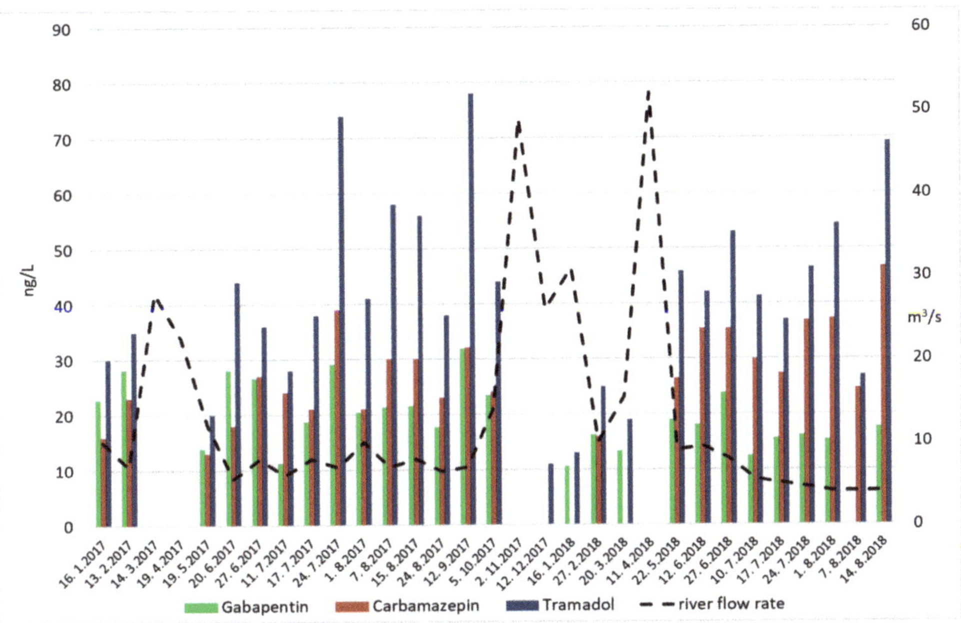

Figure 8. Changes in concentrations of selected PPCPs on profile No. 5 in relation to water discharge.

3.3. Removal of PPCP Substances during the Production of Drinking Water Using Artificial Recharge

The results of artificial recharge clearly show that the efficacy of PPCP removal is relatively high. A total of 26 substances were detected on the No. 5 profile Sojovice weir and 16 of them completely disappeared after treatment. This process does not involve mechanical treatment, which does not affect the monitored substances. Therefore, their removal takes place only when the surface water has passed and been infiltrated through the unconsolidated rock.

Only six substances occur systematically in groundwater tapped from the system of large-diameter wells in the vicinity of infiltration ponds. They include, in particular, acesulfame and oxypurinol at concentrations exceeding 100 ng/L that were evidently associated with the wastewater treatment plant in Mladá Boleslav. Other substances, which systematically occur in the produced water, were only at concentrations of the first tens of ng/L. These include carbamazepine, sulfamethoxazole, primidone, and lamotrigine.

Figure 9 gives the average values of ibuprofen and paraxanthine. These substances, however, occur rather rarely in water from artificial recharge (ibuprofen two times and gabapentin five times) obviously without exhibiting any systematic trend.

Water tapped along the infiltration ponds is a mixture of Quaternary groundwater and water from infiltration technology. The model study by Milický [23] assumes the dominant proportion (70–80%) of the component comes from artificial recharge. In September 2018, in order to verify the influence of groundwater inflow, the "natural background" was checked on a one-time basis in monitoring boreholes of the Káraný waterworks east of the artificial recharge objects. The results showed that PPCPs that systematically occurred in a mixed sample of collected water did not appear in any of the boreholes, as shown in Table 1. Consequently, these micropollutants must have originated from the river water. On the contrary, ibuprofen—only randomly occurring in the mixed sample—was detected in the boreholes. These results indicate the existence of another source of contamination independent of the river water that that spreads in the quaternary aquifer from the east.

Drinking water production in the Káraný waterworks varies depending on the demand. When the infiltration is reduced, then the influence of river water on total chemistry is reduced. The resulting concentrations of PPCPs can therefore be affected by the Quaternary aquifer. However, to prove this assumption, it would be necessary to monitor boreholes, and to have a longer series of analyses available.

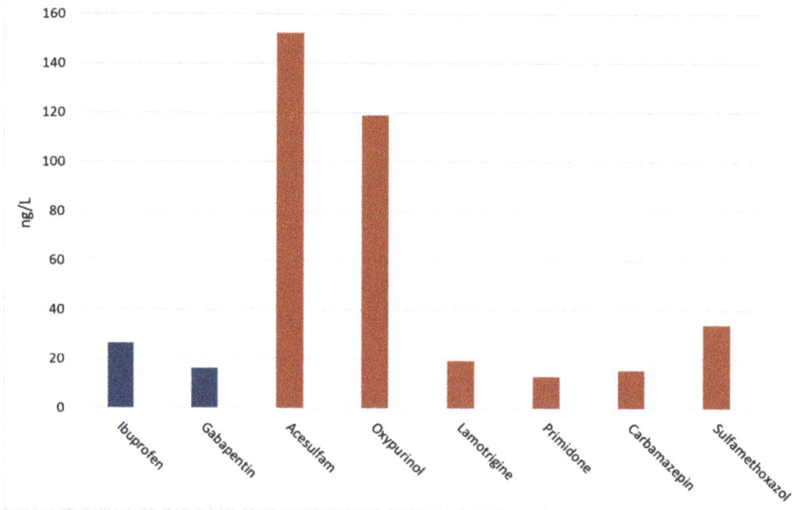

Figure 9. Long-term average values of PPCPs detected in water after artificial recharge (blue are mean values of random occurrences, red are mean values of systematically occurring substances).

Table 1. Contents of PPCP substances in monitoring wells SO 1032, HS1, HS2, HS3, HS4 and HS5 of the Káraný waterworks (position of well depicted in Figure 1, data in bold are higher than detection limit).

		Well					
Name of Monitoring Well		**SO 1032**	**HS1**	**HS2**	**HS3**	**HS4**	**HS5**
PPCP	**Unit**						
Bisfenol A	**ng/L**	**63.4**	<50.0	**217**	<50.0	<50.0	**85.3**
Ibuprofen	**ng/L**	<20.0	<20.0	**34.8**	**25.9**	<20.0	<20.0
Caffein	**ng/L**	<100	<100	**186**	**128**	<100	<100
Ketoprofen	**ng/L**	<10.0	<10.0	**24.5**	**53.1**	**20.2**	<10.0
Saccharin	**ng/L**	<50.0	<50.0	**51.5**	<50.0	<50.0	**81.9**
Paracetamol	**ng/L**	**10.4**	<10.0	<10.0	<10.0	<10.0	**138**
Paraxanthine	**ng/L**	<100	<100	**114**	**172**	<100	<100
Bisfenol S	**ng/L**	<50.0	<50.0	**4570**	**565**	<50.0	**62.8**

3.4. Removal of PPCPs during Drinking Water Production Using Bank Infiltration Technology

The water quality of bank infiltration was studied at the Káraný waterworks on a mixed sample from 685 wells. Not all wells for bank infiltration were always contributing to the resulting mixed sample, as some were often disconnected. The analyses from No. 3 and No. 4 profiles represent point data, whereas the chemical composition of the water from the bank infiltration characterizes the average concentrations over the entire length of the bank profile. An overview of PPCPs that occurred in a mixture of bank infiltration is given in Table 2. The results show that no substances systematically occurring in this water were found.

Acesulfame occurred most frequently (in 6 cases of the 31 analyzed samples), while other substances were detected at most three times. Caffeine showed the highest values in the first hundreds of ng/L, while paraxanthine was found in only one case. The amounts of all other substances were in tens of ng/L. All micropollutants were found to behave independently. There is no pair that shows a common trend.

Table 2. An overview of PPCPs in a mixed sample of bank infiltration (the table shows only the observed time periods when at least one substance was found above the detection limit).

PPCP	Unit	Sampling Data													
		16.1.2017	19.5.2017	20.6.2017	27.6.2017	11.7.2017	24.8.2017	5.10.2017	2.11.2017	12.12.2017	11.4.2017	22.5.2017	27.6.2017	10.7.2017	14.8.2017
Ibuprofen	ng/L	54					31								
Diclofenac	ng/L			31											
Caffein	ng/L					140			230					148	
Chloramphenicol	ng/L										32				
Saccharin	ng/L		65												
Gabapentin	ng/L				11										
Paracetamol	ng/L				10	16									
Clarithromycin	ng/L														
Roxithromycin	ng/L														
Paraxanthine	ng/L					141									
Acesulfam	ng/L							57	64			58	60	60	59
Oxypurinol	ng/L									72				50	61
Primidone	ng/L														11

4. Discussion

Two-year monitoring results demonstrated a wide range of PPCPs in the Jizera River basin. The main cause of their spread is the low efficiency of sewage treatment plants and the presence of a significant source of PPCPs—A psychiatric hospital at Kosmonosy.

When comparing the two purification technologies used in the Káraný waterworks, the bank infiltration as a process eliminating the pharmaceuticals from the original river water was found more efficient relative to artificial recharge technology. The reason for this may be the higher content of clay minerals in the bottom and sides of the river bed, while artificial recharge takes place in clean sands and gravel.

The issues of the occurrence of PPCPs in drinking water in the Káraný waterworks is more or less associated only with artificial recharge. The problem can be solved in several ways. The cheapest is the optimization of drinking water production. The waterworks would limit the artificial recharge operation in low flow periods, when PPCPs in the Jizera River water are high. The second option is to equip the wastewater treatment plant in Mladá Boleslav with an active carbon filter. The same filter in the pretreatment technology in Káraný waterworks should solve the problem.

5. Conclusions

- Raw water from the Jizera River contains a range of PPCPs in concentrations ranging from nanograms to micrograms per liter.
- The wastewater treatment plant at Mladá Boleslav significantly affects the quality of water used for production of drinking water in the Káraný waterworks. Of the 44 substances systematically detected in the river water below the wastewater treatment plant, 18 substances showed elevated content. The increase of telmisartan and iomeprol concentrations was approximately 100 ng/L, and in the case of oxypurinol, nearly 200 ng/L.
- The water discharge during flood periods significantly affects the time-related variability in PPCP content in river water. At elevated water discharges (10 m^3/s), PPCP concentrations were always below the detection limit of the analytical method used.
- The time-related variability of some PPCPs in river water during 24 h demonstrates the need for a uniform time schedule for sampling.
- Acesulfame and oxypurinol were detected in concentrations exceeding 100 ng/L in purified water using artificial recharge technology. Both these substances originate from a wastewater treatment plant comprising waste water from the psychiatric hospital at Kosmonosy. Systematic occurrence of carbamazepine, sulfamethoxazole, primidone, and lamotrigine in amounts of the first tens of ng/L originated from the river water used for artificial recharge.
- Ibuprofen and gabapentin were detected at irregular time intervals in drinking water produced through artificial recharge. Ibuprofen may come from the environment of the Quaternary aquifer when the share of artificial recharge on the total balance of mixed sample is lower.
- Bank infiltration is a technology that removes PPCPs in a more effective way than artificial recharge. None of the monitored substances occurred systematically in the mixed sample. Acesulfame occurred most frequently (in 6 cases of the 31 samples analyzed), while other substances were detected three times in maximum in concentrations of only the first tens of ng/L. The occurrences of individual detected substances were not correlated.

Author Contributions: Z.H. and A.H. conceived and designed the experiments; P.E., D.R., and E.N. performed the monitoring; Z.H. analyzed the data; P.E., D.R., and E.N. contributed materials/analysis tools; Z.H. wrote the paper.

Funding: This research was funded by the project Water for Prague CZ.07.1.02/0.0/0.0/16_023/0000118.

Conflicts of Interest: The authors declare no conflict of interest. The founding sponsor had no role in the design of the study; in the collection, analyses, or interpretation of data; in the writing of the manuscript, or in the decision to publish the findings.

References

1. Zhang, Z.L.; Zhou, J.L. Simultaneous determination of various pharmaceutical compounds in water by solid-phase extraction-liquid chromatography-tandem mass spectrometry. *J. Chromatogr.* **2007**, *1154*, 205–213. [CrossRef] [PubMed]
2. Huerta-Fontela, M.; Galceran, M.T.; Ventura, F. Fast liquid chromatography-quadrupole-linear ion trap mass spectrometry for the analysis of pharmaceuticals and hormones in water resources. *J. Chromatogr.* **2010**, *1217*, 4212–4222. [CrossRef] [PubMed]
3. Ferrer, I.; Thurman, E.M. Analysis of 100 pharmaceuticals and their degradates in water samples by liquid chromatography/quadrupole time-of-flight mass spectrometry. *J. Chromatogr.* **2012**, *1259*, 148–157. [CrossRef] [PubMed]
4. Richardson, S.D. Environmental mass spectrometry: Emerging contaminants and current issues. *Anal. Chem.* **2006**, *78*, 4021–4045. [CrossRef] [PubMed]
5. Rozman, D.; Hrkal, Z.; Eckhardt, P.; Novotna, E.; Boukalova, Z. Pharmaceuticals in groundwaters: A case study of the psychiatric hospital at Horní Beřkovice, Czech Republic. *Environ. Earth Sci.* **2015**, *73*, 3775–3784. [CrossRef]

6. Rozman, D.; Hrkal, Z.; Váňa, M.; Vymazal, J.; Boukalová, Z. Occurrence of Pharmaceuticals in Wastewater and Their Interaction with Shallow Aquifers: A Case Study of Horní Beřkovice, Czech Republic. *Water* **2017**, *9*, 218. [CrossRef]
7. Chena, Y.; Vymazal, J.; Březinová, T.; Koželuch, M.; Kulec, L.; Huangd, J.; Chena, Z. Occurrence, removal and environmental risk assessment of pharmaceuticals and personal care products in rural wastewater treatment wetlands. *Sci. Total Environ.* **2016**, *566–567*, 1660–1669. [CrossRef] [PubMed]
8. Vymazal, J.; Dvořáková Březinová, T. Removal of saccharin from municipal sewage: The first results from constructed wetlands. *Chem. Eng. J.* **2016**, *306*, 1067–1070. [CrossRef]
9. Vymazal, J.; Dvořáková Březinová, T.; Koželuh, M.; Kuleb, L. Occurrence and removal of pharmaceuticals in four full-scale constructed wetlands in the Czech Republic—The first year of monitoring. *Ecol. Eng.* **2017**, *98*, 354–364. [CrossRef]
10. Lubliner, B.; Redding, M.; Ragsdale, D. *Pharmaceuticals and Personal Care Products in Municipal Wastewater and Their Removal by Nutrient Treatment Technologies*; Washington State Department of Ecology: Olympia, WA, USA, 2010; Volume 230.
11. Kasprzyk-Hordern, B.; Dinsdale, R.M.; Guwy, A.J. The removal of pharmaceuticals, personal care products, endocrine disruptors and illicit drugs during wastewater treatment and its impact on the quality of receiving waters. *Water Res.* **2009**, *43*, 363–380. [CrossRef] [PubMed]
12. Ternes, T.A. Occurrence of drugs in German sewage treatment plants and rivers. *Water Res.* **1998**, *32*, 3245–3260. [CrossRef]
13. Tauxe-Wuersch, A.; De Alencastro, L.F.; Grandjean, D.; Tarradellas, J. Occurrence of several acidic drugs in sewage treatment plants in Switzerland and risk assessment. *Water Res.* **2005**, *39*, 1761–1772. [CrossRef] [PubMed]
14. Rodriguez, E.; Campinas, M.; Acero, J.L. Investigating PPCP Removal from Wastewater by Powdered Activated Carbon/Ultrafiltration. *Water Air Soil Pollut.* **2016**, *227*, 177. [CrossRef]
15. Rivera-Utrilla, J.; Sanchez-Polo, M.; Ferro-Garcia, M.A.; Prados-Joya, G.; Ocampo-Perez, R. Pharmaceuticals as emerging contaminants and their removal from water. A review. *Chemosphere* **2013**, *93*, 1268–1287. [CrossRef] [PubMed]
16. Jones, O.A.; Lester, J.N.; Voulvoulis, N. Pharmaceuticals: A threat to drinking water? *Trends Biotechnol.* **2005**, *23*, 163–167. [CrossRef] [PubMed]
17. Stackelberg, P.E.; Gibs, J.; Furlong, E.T.; Meyer, M.T.; Zaugg, S.D.; Lippincott, R.L. Efficiency of conventional drinking-water-treatment processes in removal of pharmaceuticals and other organic compounds. *Sci. Total Environ.* **2007**, *377*, 255–272. [CrossRef] [PubMed]
18. Kožíšek, F.; Pomykačová, I.; Jeligová, H.; Čadek, V.; Svobodova, V. Survey of human pharmaceuticals in drinking water in the Czech Republic. *J. Water Health* **2013**, *11*, 84–97. [CrossRef] [PubMed]
19. Godoy, A.A.; Kummrow, F.; Pamplin, P.A.Z. Occurrence, ecotoxicological effects and risk assessment of antihypertensive pharmaceutical residues in the aquatic environment—A review. *Chemosphere* **2015**, *138*, 281–291. [CrossRef] [PubMed]
20. Commission Implementation Decision (EU) 2015/495 of 2015 Establishing a watch list of substances for Union-wide monitoring in the field of water policy pursuant to Directive 2008/105/EC of the European Parliament and of the Council. *Off. J. Eur. Union* **2015**, *56*, L 78/40.
21. Jedlička, B.; Kněžek, M. Artificial recharge. *Hydrogeol. Roč* **1968**, *1*, 131–148. (in Czech).
22. Skalický, M. Artificial recharge of the Káraný site as a tool for solving the lack of groundwater for water use (in Czech). In Proceedings of the "Groundwater in Water Treatment 2015", Pardubice, Czech Republic, 1–2 April 2015.
23. Milický, M. *Káraný Waterwork—Evaluation of the Development of the Groundwater Resources and Quality of Groundwater during the Hydrological Year 2017, Optimization of the Operation of the Artificial Infiltration Complex, Optimization of the Operation of Sources of Bank Infiltration with Increased Nitrate Content (in Czech)*; Final Rapport; PROGEO Company: Roztoky, Czech Republic, 2018.

Article

Long-Term River Water Quality Trends and Pollution Source Apportionment in Taiwan

Marsha Savira Agatha Putri [1], Chao-Hsun Lou [1], Mat Syai'in [2], Shang-Hsin Ou [3] and Yu-Chun Wang [1,*]

1 Department of Environmental Engineering, College of Engineering, Chung Yuan Christian University, Taoyuan 32023, Taiwan; marshasavira20@gmail.com (M.S.A.P.); louislou9333@gmail.com (C.-H.L.)
2 Department of Automation Engineering, Shipbuilding Institute of Polytechnic Surabaya, East Java 60111, Indonesia; matt.syaiin@ppns.ac.id
3 Taiwan Water Corporation, Taichung 32404, Taiwan; shou24@cycu.org.tw
* Correspondence: ycwang@cycu.edu.tw; Tel.: +886-3-2654916

Received: 23 August 2018; Accepted: 4 October 2018; Published: 8 October 2018

Abstract: The application of multivariate statistical techniques including cluster analysis and principal component analysis-multiple linear regression (PCA-MLR) was successfully used to classify the river pollution level in Taiwan and identify possible pollution sources. Water quality and heavy metal monitoring data from the Taiwan Environmental Protection Administration (EPA) was evaluated for 14 major rivers in four regions of Taiwan with the Erren River classified as the most polluted river in the country. Biochemical oxygen demand (6.1 ± 2.38), ammonia (3.48 ± 3.23), and total phosphate (0.65 ± 0.38) mg/L concentration in this river was the highest of the 14 rivers evaluated. In addition, heavy metal levels in the following rivers exceeded the Taiwan EPA standard limit (lead: 0.01, copper: 0.03, and manganese: 0.03) mg/L concentration: lead-in the Dongshan (0.02 ± 0.09), Jhuoshuei (0.03 ± 0.03), and Xinhuwei Rivers (0.02 ± 0.02) mg/L; copper: in the Dahan (0.036 ± 0.097), Laojie (0.06 ± 1.77), and Erren Rivers are (0.05 ± 0.158) mg/L; manganese: in all rivers. A total 72% of the water pollution in the Erren River was estimated to originate from industrial sources, 16% from domestic black water, and 12% from natural sources and runoff from other tributaries. Our research demonstrated that applying PCA-MLR and cluster analysis on long-term monitoring water quality would provide integrated information for river water pollution management and future policy making.

Keywords: Taiwan rivers; water quality; multivariate statistical analysis; river pollution index; pollution source apportionment

1. Introduction

Surface water quality is a matter of critical concern in developing countries because of growing population, rapid industrialization, urbanization, and agricultural modernization [1]. Of all water bodies, rivers are the most vulnerable to pollution because of their role in carrying agricultural run-off and municipal and industrial wastewater [2]. Water quality experts and decision makers are confronted with significant challenges in their efforts to manage surface water resources due to these complex issues [3]. Spatial variation and source apportionment characterization of water quality parameters can provide a detailed understanding of environmental conditions and help researchers to establish priorities for sustainable water management [4].

In recent years, the national income and standard of living in Taiwan have considerably improved following the nation's focus on economic development [5]. Rapid industrial development in Taiwan, including increased vehicle use, electrical power generation, and manufacturing of food, beverages, textiles, plastic, and metal, has affected pollution levels and other environmental problems, specifically water pollution [6]. In 1998, the Taiwan Environmental Protection Administration (EPA) reported that 16% (2088 km) of the total length of Taiwan's 21 major rivers was ranked as severely polluted, while another 22% were considered lightly and moderately polluted [7].

The Taiwan EPA uses the river pollution index (RPI) to explore monitoring trends for both long-term planning and day-to-day management of surface water quality. The RPI involves four parameters: dissolved oxygen (DO), biochemical oxygen demand (BOD), suspended solids (SS), and ammonia nitrogen (NH_3-N). The overall index is divided into four pollution levels as follow: non-polluted, lightly polluted, moderately polluted, and severely polluted [8]. Previous research has used the RPI to evaluate the pollution levels of the following rivers: the Tanshui River [9,10], Kaoping River [11], Chuo-shui River, Beigang River, Jishui River, Agongdian River, and Sichong River [12] in Taiwan, and the Mahmoudia Canal in Egypt [13].

The application of multivariate statistical analysis for cluster analysis, principal component analysis (PCA), and source apportionment by multiple regression on principal components provides a detailed understanding of water quality and the ecological status of the studied systems for improved interpretation of these complex data matrices [14,15]. Such analyses also facilitate the identification of possible pollution sources that affect the water systems and offers a valuable tool for reliable management of water resources and the determination of potential solutions to pollution problems.

As shown in Figure 1, this study was conducted in three phases with three main objectives: (1) evaluate and compare water quality and heavy metal data of 14 major Taiwan rivers with the Taiwan EPA standards; (2) classify the contamination level of those 14 major Taiwan rivers then determine the most polluted river; and (3) identify the major possible pollution source apportionment affecting water quality in the most polluted river.

2. Materials and Methods

2.1. Study Area and Data Collection

The subtropical island of Taiwan has 151 major and minor rivers with a combined total length of 3717 km. Most rivers flow down from high mountains in short and steep courses [16]. Figure 2 displays the 14 major rivers that were selected for analysis of their water quality and heavy metal concentrations: Dahan, Danshuei, Jilong, and Laojie rivers in Northern Taiwan; Dongshan River in Eastern Taiwan; Jhuoshuei, Wu, and Xinhuwei rivers in Central Taiwan; and Erren, Gaoping, Jishuei, Puzi, Yanshuei, and Zengwun rivers in Southern Taiwan. These 14 rivers were selected because those rivers are the biggest in each area and the most polluted rivers according to the previous Taiwan EPA report.

Table 1 shows information on the 14 major rivers and monitoring stations. Water quality data from 2002 until 2016 were provided by Taiwan EPA for each monitoring site in winter (December–February), spring (March–May), summer (June–August), and fall (September–November). The sampling procedures were conducted according to standard operational procedures summarized in Supplementary Table S1.

The 14 water quality parameters seasonally measured included water temperature, air temperature, RPI, pH, suspended solid (SS), BOD, DO, ammonia, conductivity, chemical oxygen demand (COD), total phosphate (TP), total organic carbon (TOC), nitrite, and nitrate. Moreover, 10 heavy metal parameters were measured: lead (Pb), arsenic (As), cadmium (Cd), chromium (Cr), zinc (Zn), mercury (Hg), Cu, Mn, silver (Ag), and selenium (Se). The unobserved data of all water quality parameters and heavy metal was treated using the random forest algorithm method [17].

2.2. Statistical Methods

Cluster analysis was used to classify the rivers based on the RPI data. Pearson's correlation analysis was used to test for linear correlation between the RPI and water quality parameters (water temperature, air temperature, conductivity, nitrate, SS, DO, BOD, COD, ammonia, TP, and TOC).

PCA-MLR was used to determine source apportionment. These three multivariate analyses were performed using SPSS 22.0 for Windows (IBM Corp., Armonk, NY, USA, 2013).

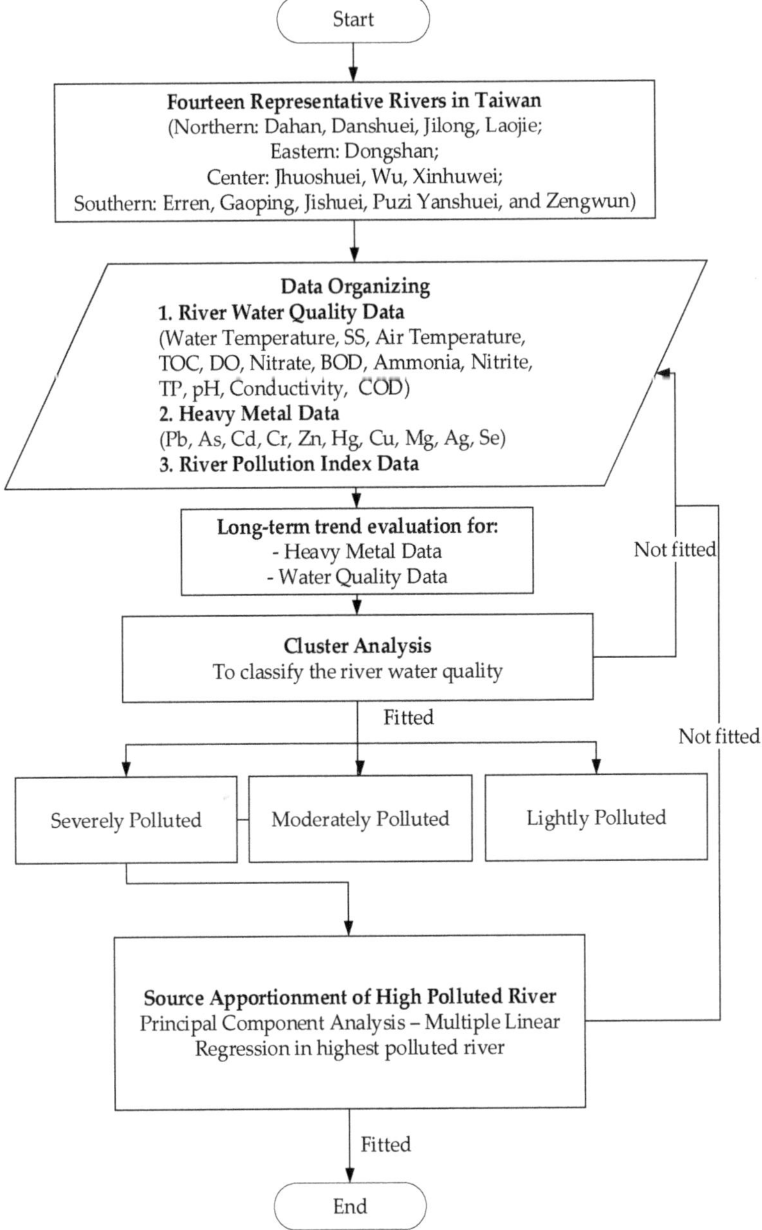

Figure 1. Flowchart of this study.

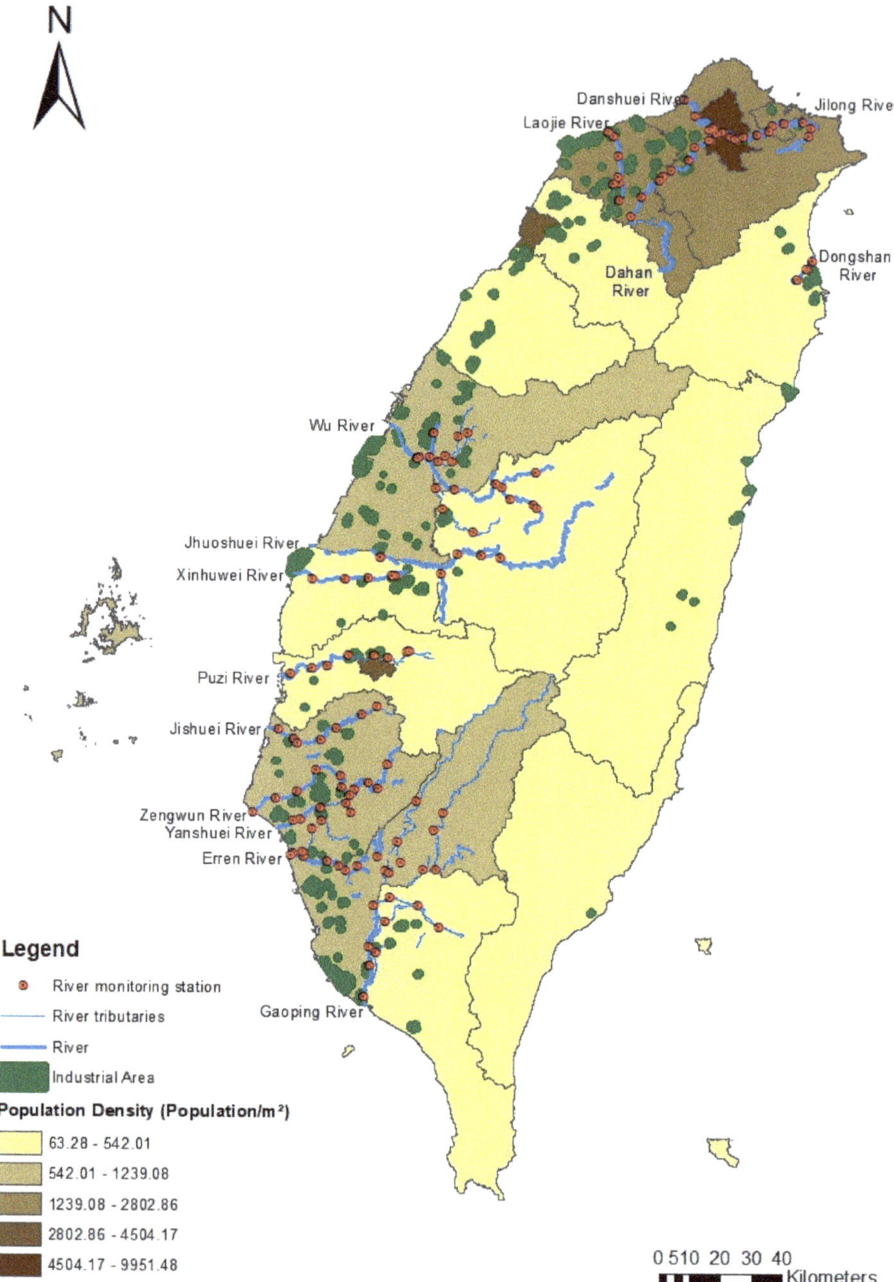

Figure 2. Study area of 14 Taiwan representative rivers and their sampling points, population density, and industrial area map.

Table 1. Representative river water monitoring sites.

Code	River Name *	Main River Length ** (km)	Average Discharge ** (m^3/s)	Number of Monitoring Stations *	Coordinate *	Municipality *
N1	Dahan	135	59.62	7	25°2′33.36″ N, 121°29′2.4″ E	New Taipei and Taoyuan
N2	Danshuei	158.7	62.96	4	25°10′30″ N, 121°24′30″ E	New Taipei and Taipei
N3	Jilong	96	39.5	13	25°6′43.92″ N, 121°27′50.4″ E	Keelung, New Taipei, and Taipei
N4	Laojie	36.7	Not available	7	24°52′21″ N, 121°13′17.4″ E	Taoyuan
E1	Dongshan	24	Not available	3	24°40′3.1″ N, 121°48′41.1″ E	Yilan
C1	Jhuoshuei	186.6	143.73	5	23°50′22″ N, 120°15′14″ E	Nantou and Yunlin
C2	Wu	119.13	55.86	18	24° 10′11.5″ N, 120°31′5.7″ E	Nantou, Taichung, and Zhanghua
C3	Xinhuwei	49.85	Not available	5	23°44′58.95″ N, 120°30′58.69″ E	Yunlin
S1	Erren	61.20	8.43	18	22°54′46.8″ N, 120°10′33.6″ E	Kaohsiung and Tainan
S2	Gaoping	171	216.96	9	22°28′59″ N, 120°34′47″ E	Kaohsiung and Pingdong
S3	Jishuei	65	9.43	7	23°17′52.8″ N, 120°6′18″ E	Tainan
S4	Puzi	75.87	13.98	8	23°30′18.52″ N, 120°29′51.53″ E	Chiayi
S5	Yanshuei	41.3	5.92	11	23° 0′14.4″ N, 120°9′0″ E	Tainan
S6	Zengwun	138.47	9.2	11	23°3′0″ N, 120°4′1.2″ E	Tainan

*: Environmental Protection Administration Executive Yuan, R.O.C (Taiwan); **: Water Resource Agency, Ministry of Economic Affair R.O.C (Taiwan), N: Northern River; E: Eastern River; C: Center River; S: Southern River.

2.2.1. Cluster Analysis

Hierarchical agglomerative cluster analysis was performed on the normalized RPI data set by Ward's method, using squared Euclidean distances as a measure of similarity. Ward's method uses an analysis of variance approach to evaluate the distances between clusters in an attempt to minimize the sum squares of any two clusters that can be formed at each step. To standardize the linkage distance represented on the y-axis, the linkage distance is reported as D_{link}/D_{max}, where D_{link} represents the quotient between the linkage distance for a particular case and D_{max} is the maximal distance multiplied by 100 [18–20]. The data used for cluster analysis are water quality data of 14 rivers from 2002 to 2016.

2.2.2. Source Apportionment Analysis

Principal component analysis (PCA) is a dimension-reduction technique that provides information on the most significant factors with a simple representation of the data. It is generally used for data structure determination and for obtaining qualitative information about potential pollution sources. However, if used alone, PCA cannot determine the quantitative contributions of the identified pollution sources in each variable [21]. Correlation analysis using Pearson's analysis was thus employed to select the high significant correlation ($r > 0.3$ with $p < 0.01$) between RPI and water quality parameters that would be inputted into PCA. The $r > 0.3$ indicates strong correlation among parameters and $p < 0.01$ signs the highly significant result [22].

Kolmogorov–Smirnov (K–S) statistics were used to test the goodness-of-fit of the data to log-normal distribution. To examine the suitability of the data for PCA, the Kaiser-Meyer-Olkin (KMO) and Bartlett's Sphericity tests were applied on the prepared dataset. In the KMO test, a value closer to 1 indicates high validity while a value <0.7 indicates an invalid analysis. Bartlett's Sphericity test was used to check the null hypothesis that the inter-correlation matrix comes from a population in which the variables are uncorrelated. For this study, the null hypothesis was rejected due to a significance level >0.01 [23]. These two tests required the selected water quality data to be fitted before PCA interpretation. If these two tests fulfil the requirement, we need to consider the total variance while a value >60%. Rotated variables with factor loading >0.7 are considered relevant and indicate a possible emission source. Next, multiple linear regression (MLR) was applied to determine the percentage of contribution for each pollution source [21,24]. In linear regression, the sum of each parameter standardization was defined as a dependent variable and the absolute principal component score as an independent variable.

3. Results

3.1. Evaluation of River Water Quality and Heavy Metal Data

Figure 3 shows the RPI dataset stacked plot from 2002 to 2016 for the 14 rivers. Between 2002 and 2016, for the four rivers in Northern Taiwan, on average 15% were severely polluted, 60% were moderately polluted, and 25% were lightly polluted. The average percentages for the rivers in Eastern Taiwan were 14% moderately polluted and 86% lightly polluted. The average percentages for the three rivers in Central Taiwan were 49% moderately and 51% lightly polluted. The average percentages for the six rivers in Southern Taiwan were 18% severely polluted, 59% moderately polluted, and 23% lightly polluted. In 2016, 65% of the total number of rivers in Taiwan were classified as moderately polluted, while the remaining 35% were classified as lightly polluted. Therefore, according to the average percentages above, from 2002–2016, the rivers in Northern, Eastern, Central, and Southern Taiwan had moderate, highest, moderate, and lowest water quality levels, respectively.

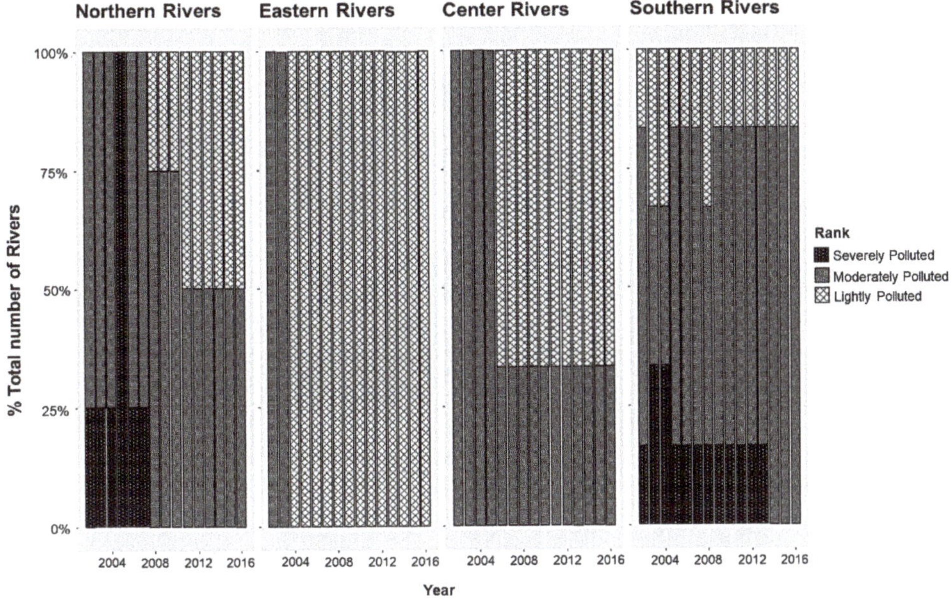

Figure 3. Long-term of RPI rank in 14 Taiwan representative rivers.

Water quality and heavy metal evaluation in each river were interpreted using the average of the monitoring sampling station data in one river. Figure 4 displays a boxplot of river water quality variables for the 14 rivers. Each boxplot displays the six-number summary of a set of data, including the minimum, first quartile, median, third quartile, maximum, and outliers' value. Overall, the central and southern rivers in Taiwan had the highest levels of pH, DO, and SS, while BOD, ammonia, and TP were the highest in the Erren River.

Heavy metal concentrations for the 14 rivers are summarized in Table 2 which displays total sample (*N*), long-term average value (mean), and standard deviation (SD). The overall mean concentrations of heavy metals were found in the following order: Hg < As < Cd < Se < Ag < Cr < Pb < Cu < Zn < Mn. In addition, the highest concentrations of heavy metals, all of which exceeded the Taiwan EPA standard limit (Table S2), were found in the following rivers: Pb in the Dongshan, Jhuoshuei, and Xinhuwei Rivers; Cu in the Dahan, Laojie, and Erren Rivers; and Mn in all rivers.

Table 2. Descriptive statistic of heavy metals in 14 Taiwan representative rivers between 2002 and 2016.

River Name	Pb (mg/L)			As (mg/L)			Cd (mg/L)			Cr (mg/L)			Zn (mg/L)		
	N	Mean	SD	*N*	Mean	SD	*N*	Mean	SD	*N*	Mean	SD	*N*	Mean	SD
Dahan	412	0.01	0.02	412	0.002	0.004	412	0.0013	0.0027	412	0.004	0.003	412	0.059	0.13
Danshuei	236	0.01	0.01	236	0.002	0.002	236	0.001	0.0003	236	0.003	0.002	236	0.031	0.03
Jilong	301	0.01	0.01	301	0.004	0.003	301	0.0011	0.0008	301	0.004	0.003	301	0.021	0.02
Laojie	767	0.01	0.03	765	0.001	0.001	767	0.0011	0.0013	765	0.003	0.003	767	0.03	0.04
Dongshan	413	0.02	0.09	413	0.004	0.006	413	0.0013	0.0028	413	0.008	0.054	413	0.187	0.34
Jhuoshuei	267	0.03	0.03	267	0.003	0.002	267	0.0013	0.001	267	0.004	0.004	267	0.079	0.09
Wu	1042	0.01	0.04	1040	0.001	0.001	1042	0.0011	0.0006	1041	0.007	0.019	1042	0.04	0.08
Xinhuwei	300	0.02	0.02	300	0.005	0.004	300	0.0011	0.0005	300	0.004	0.004	300	0.055	0.06
Erren	668	0.02	0.02	668	0.008	0.005	668	0.0019	0.0026	668	0.009	0.074	668	0.125	0.28
Gaoping	948	0.02	0.02	944	0.003	0.005	948	0.001	0.0003	944	0.004	0.006	948	0.048	0.29
Jishuei	439	0.01	0.01	439	0.009	0.005	439	0.001	0.0003	439	0.004	0.006	438	0.022	0.03
Puzi	442	0.01	0.01	442	0.005	0.003	442	0.001	0.0003	442	0.005	0.01	442	0.065	0.31
Yanshuei	343	0.01	0.01	343	0.009	0.005	343	0.0011	0.001	343	0.005	0.012	343	0.04	0.07
Zengwun	665	0.01	0.02	665	0.003	0.003	665	0.001	0.0005	665	0.004	0.004	665	0.023	0.04

River Name	Hg (mg/L)			Cu (mg/L)			Mn (mg/L)			Ag (mg/L)			Se (mg/L)		
	N	Mean	SD	*N*	Mean	SD	*N*	Mean	SD	*N*	Mean	SD	*N*	Mean	SD
Dahan	384	0.0004	0.00038	412	0.036	0.097	412	0.22	0.62	412	0.0026	0.0027	307	0.0013	0.00045
Danshuei	220	0.00044	0.00051	236	0.018	0.032	236	0.12	0.12	236	0.0028	0.0043	176	0.0013	0.00045
Jilong	715	0.00038	0.0003	767	0.007	0.011	765	0.11	0.12	765	0.0028	0.0049	570	0.0013	0.00044
Laojie	385	0.00032	0.00015	413	0.06	1.777	413	0.22	0.7	413	0.0024	0.002	308	0.0013	0.00045
Dongshan	285	0.00033	0.00012	301	0.005	0.009	301	0.1	0.06	301	0.0029	0.003	252	0.0013	0.00045
Jhuoshuei	251	0.00041	0.00036	267	0.019	0.03	267	0.69	1.16	267	0.0023	0.0017	191	0.0013	0.00049
Wu	970	0.00031	0.00006	1042	0.009	0.012	1040	0.11	0.17	1040	0.0023	0.0019	759	0.0013	0.0006
Xinhuwei	280	0.00034	0.00012	300	0.012	0.019	300	0.36	0.47	300	0.0022	0.0017	220	0.0013	0.00051
Erren	632	0.00058	0.00017	668	0.05	0.158	668	0.23	0.24	668	0.0025	0.0033	519	0.0014	0.00087
Gaoping	879	0.00034	0.00018	948	0.009	0.016	944	0.41	0.71	944	0.0028	0.0076	689	0.0013	0.00057
Jishuei	411	0.00034	0.00017	439	0.005	0.004	439	0.25	0.18	439	0.0024	0.0025	330	0.0013	0.00044
Puzi	415	0.00032	0.00013	442	0.007	0.043	442	0.23	0.24	442	0.0023	0.0022	327	0.0017	0.00197
Yanshuei	319	0.00032	0.0001	343	0.011	0.016	343	0.28	0.19	343	0.0023	0.0017	256	0.0013	0.00045
Zengwun	622	0.00034	0.00016	665	0.006	0.009	665	0.18	0.52	665	0.0023	0.002	490	0.0013	0.00044

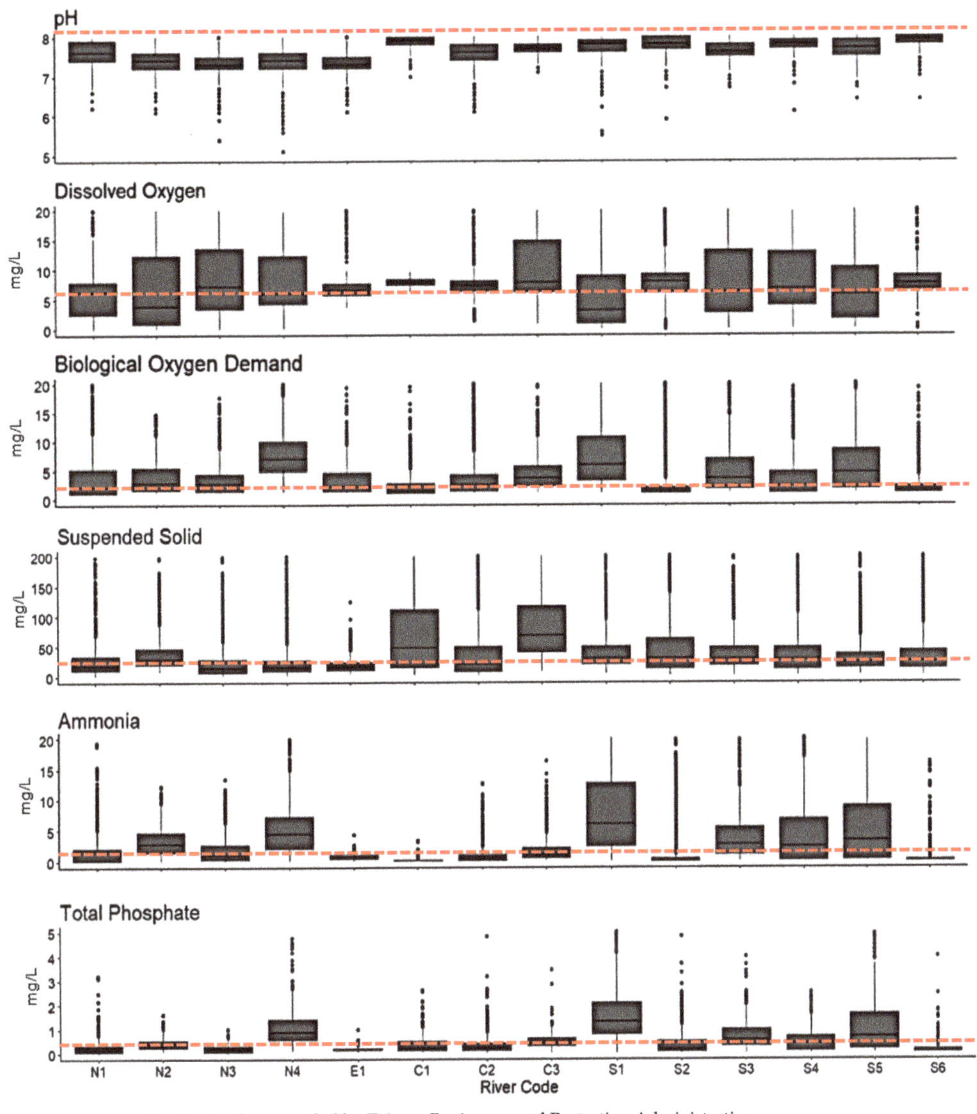

Figure 4. Boxplot of river water quality parameters for 14 representative rivers in Taiwan (see Table 1 for the river names associated with the river codes).

3.2. Cluster Analysis

Figure 5 shows the result of cluster analysis of the water quality variation tendencies among the targeted monitoring sites. Three significant groups ($p < 0.01$) were classified comprehensively based on the RPI value of each river. Cluster 1, classified as lightly polluted, included the Dongshan, Wu, and Zengwun rivers. Cluster 2, classified as moderately polluted, included the Dahan, Danshuei, Jilong, Jhuoshuei, Xinhuwei, Gaoping, Jishuei, Puzi, and Yanshuei rivers. Cluster 3, classified as severely polluted, included the Erren River. These results were fit in comparison with the RPI value of each river presented in Supplementary Table S3.

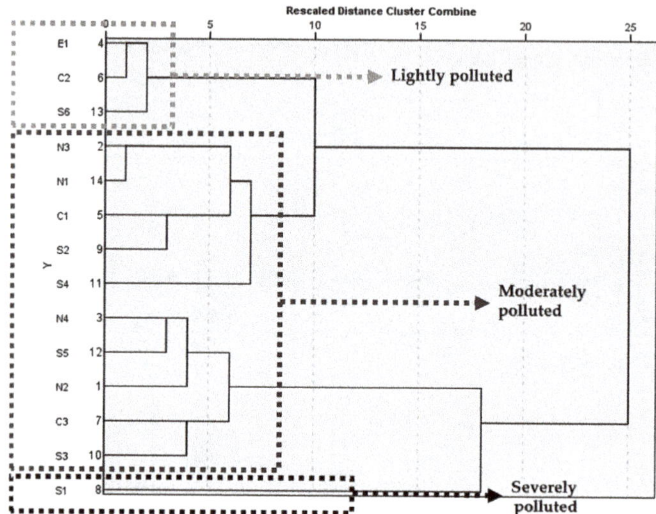

Figure 5. Dendrogram of RPI rank for 14 representative rivers in Taiwan (see Table 1 for the river names associated with the river codes).

3.3. Source Identification and Apportionment

This study analyzed source apportionment for the most polluted river, namely, the Erren River. The result of Pearson's correlation analysis is shown in Table 3. The significant water quality parameters correlated with RPI were −0.376 for conductivity, −0.719 for DO, 0.621 for BOD, 0.339 for COD, 0.512 for SS, 0.533 for coliform, 0.587 for ammonia, 0.402 for TP, 0.383 for TOC, −0.301 for nitrate, and 0.308 for nitrite. Biochemical oxygen demand, DO, SS, and ammonia were correlated with RPI because the RPI value was calculated according to the concentration of these four parameters. However, other parameters including conductivity, COD, coliform, TP, TOC, nitrate, and nitrite also had strong correlation with RPI. Therefore, these water quality parameters can be assumed to have a significant impact on river pollution levels. These significant factors were thus selected for further PCA analysis. Conductivity, likewise, showed high correlation with nitrate (0.564, $p < 0.01$) indicating that nitrate acts as an electrolyte along with organic matter in water [25].

The result indicated that PCA could be applied due to KMO's value being 0.76 and Barlett's test significance being 0.00. Figure 6 shows that there were three components in the scree plot defining the most dominant component among all variances. The three selected components in the dash mark will be defined as the three factors in the principal components which have a 62.3% total cumulative of variance percentage (Table 4). It means 62.3% variability in water quality data has been modelled by the extracted factor. Thus, it indicates that this model is properly acceptable to continue to the next step. The extracted varimax rotation of principal components among the selected parameters from correlation analysis is shown in Table 5. The bold and marked values indicate the dominant parameters in each factor.

Table 3. Correlation matrix between RPI and other water quality data in the Erren River.

Parameter	RPI	Air Temperature	Water Temperature	pH	Conductivity	DO	BOD	COD	SS	Coliform	Ammonia	TP	TOC	Nitrate	Nitrite
RPI	1	−0.089	−0.0072	−0.438	−0.376 **	−0.719 **	0.621 **	0.339 **	0.512 **	0.333 **	0.587 **	0.402 **	0.383 **	−0.301 **	0.308 **
Air Temperature	-	1	0.875 **	−0.082 **	−0.067 **	0.055 *	−0.165 **	−0.169 **	0.053 *	−0.068 **	−0.174 **	−0.313 **	−0.192 **	0.130 **	−0.006
Water Temperature	-	-	1	−0.1 **	−0.043	0.052 *	−0.0167 **	−0.172 **	0.018	−0.080 **	−0.178 **	−0.3 **	−0.137 **	0.063	−0.046
pH	-	-	-	1	0.039	0.607 **	−0.256 **	−0.144 **	0.001	−0.119 **	0.062 **	0.231 **	−0.227 **	0.121 **	−0.068
Conductivity	-	-	-	-	1	0.515 **	−0.173 **	−0.182 **	−0.561 **	−0.016	−0.145 **	−0.166 **	−0.247 **	0.564 **	0.113 **
DO	-	-	-	-	-	1	−0.415 **	−0.346 **	0.062 **	−0.069 **	−0.194 **	−0.125 **	−0.434 **	0.254 **	−0.176 **
BOD	-	-	-	-	-	-	1	0.853 **	−0.026	0.550 **	0.419 **	0.407 **	0.788 **	−0.203 **	0.098 *
COD	-	-	-	-	-	-	-	1	0.133 **	0.255 **	0.548 **	0.603 **	0.854 **	−0.150 **	0.093 *
SS	-	-	-	-	-	-	-		1	−0.011	−0.065 **	−0.021	−0.106 **	0.245 **	−0.095 *
Coliform	-	-	-	-	-	-	-		-	1	0.036	0.042	0.565 **	−0.051	−0.046
Ammonia	-	-	-	-	-	-	-		-	-	1	0.738 **	0.562 **	−0.093 *	0.098 *
TP	-	-	-	-	-	-	-		-	-	-	1	0.501 **	−0.006	0.102 *
TOC	-	-	-	-	-	-	-		-	-	-	-	1	−0.248 **	0.094
Nitrate	-	-	-	-	-	-	-		-	-	-	-	-	1	−0.066
Nitrite	-	-	-	-	-	-	-		-	-	-	-	-	-	1

*: Correlation is significant at the 0.05 level (2-tailed); **: Correlation is significant at the 0.01 level (2-tailed).

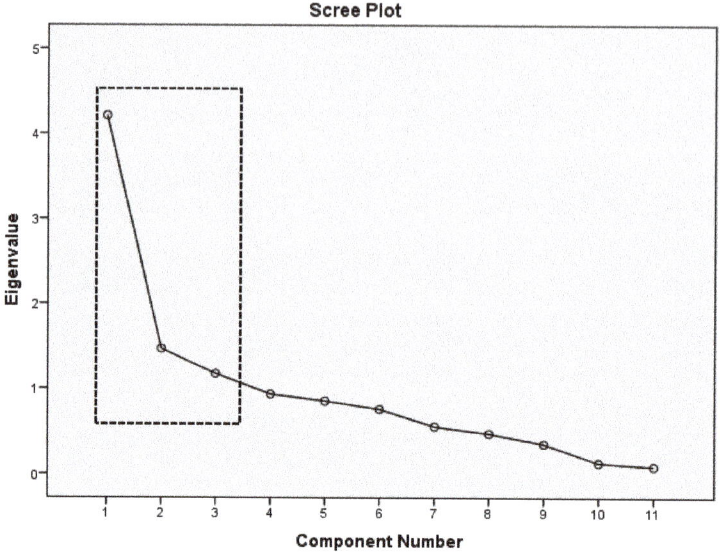

Figure 6. Scree plot of PCA.

Table 4. Initial eigenvalues and selected component loading after varimax rotation for PCA results in the Erren River.

Component (Factor)	Initial Eigenvalues		
	Total	**% of Variance**	**Cumulative %**
1	4.208	38.25	38.25
2	1.471	13.37	51.62
3	1.175	10.68	62.3

Table 5. Varimax rotated factor of PCA of Erren River water quality.

Parameter	Component (Factor)		
	1	**2**	**3**
DO	−0.050	**−0.734**	0.301
BOD	0.487	**0.748**	0.066
COD	0.564	**0.718**	0.137
Ammonia	**0.858**	0.069	−0.086
TP	**0.866**	0.027	0.005
TOC	0.671	**0.727**	0.024
Conductivity	−0.203	−0.231	**−0.737**
Nitrate	−0.097	−0.458	**0.700**
SS	−0.185	−0.008	**0.709**
Nitrite	−0.067	**0.714**	−0.558
Coliform	0.046	**0.810**	0.031

The MLR results in Table 6 show that all regression results were significant ($p < 0.01$). Factor 1, accounting for 72% of the total variance, had strong and positive loadings on ammonia and TP. High concentrations of TP and ammonia in surface water can come from various sources, including municipal and industrial effluent [14]. Factor 2, accounting for 16% of the total variance, possibly originates from domestic black water factors related to the high loadings on coliform, BOD, COD, TOC, and nitrite [26,27]. Factor 3, accounting for 12% of the total variance, had high and positive loadings on conductivity, nitrate, and SS, and thus was interpreted as a mineral component of river water or runoff precipitation [28]. The source apportionment of this study is reliable because PCA-MLR showed the good fitted receptor model (R^2 > 0.74). The contribution percentage of each factor is shown in Figure 7.

Table 6. Contribution percentage of each factor from MLR results.

Model	B	Sig.	Percentage	R^2
Constant	−0.602	0.000	-	
Regression Factor 1	2.573	0.000	72%	0.74
Regression Factor 2	0.554	0.000	16%	
Regression Factor 3	0.440	0.000	12%	

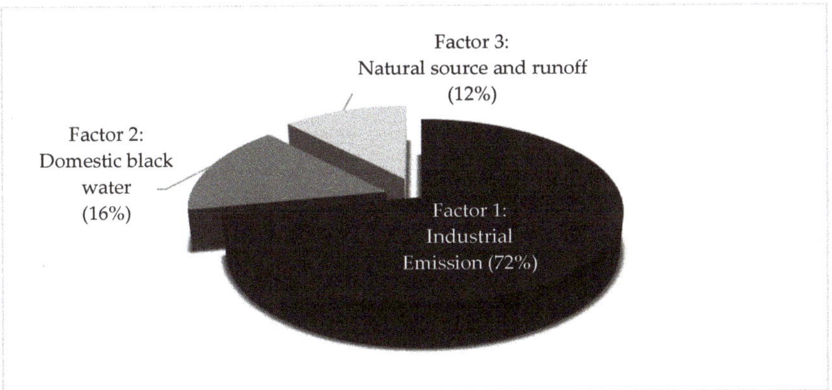

Figure 7. Source apportionment of water pollutant in the Erren River.

4. Discussion

The increasing population density in Taiwan is a significant source of domestic water pollution. Wastewater from agriculture, farming, and urban activities can also be major pollution sources causing diverse problems, such as toxic algal blooms, loss of oxygen, fish kills, loss of biodiversity (including species important for commerce and recreation), and loss of aquatic plant beds and coral reefs [26]. In addition, despite the decreasing number of domestic swine farms in Taiwan after it joined the World Trade Organization, approximately 7 million swine are still being raised in the country and their waste must be disposed of. Aside from domestic pollution sources, industrial wastewater is a major water pollution source as well. During the last three decades, Taiwan developed into a large trading economy with nearly 11,000 manufacturing plants disposing various contaminants [29]. Moreover, the location of industrial area and high population density in Taiwan is scattered. In other words, industrial and high population areas around Taiwan rivers are not only located in the downstream areas, but also in the upstream areas (see Supplementary Figure S1). Thus, water pollution in Taiwan rivers is spread along the river (see Supplementary Figure S2). We interpreted the water quality and heavy metals characteristics for each river that could be used by the Taiwan government to plan proper river management strategy.

Our study used cluster analysis on large-scale data in one country to classify the pollution level of major rivers. Previous research has shown that cluster analysis is useful for classifying rivers that have similar water quality characteristics. For example, Shrestha and Kazama 2007 [18] reported that cluster analysis results represent the influence of land use, residential sewage, agricultural activities, and industrialization, which can have major impacts on water quality. Another study grouped monitoring sites in rivers in South Florida into three groups (low, moderate, and high pollution) on the basis of their similar water quality characteristics [30].

In the current study, Erren River was determined to be the most polluted river among the other major rivers in Taiwan. The most significant water pollutants were identified to have originated from industrial activity, domestic black water, runoff from other rivers, and natural sources, including climate conditions. Given that the Erren River is located in an industrialized and urbanized area [31], the level of water pollutants in the river is very high due to huge amounts of nutrient salts, including organic pollutants, ammonia, and total phosphate. These pollutants are associated with possible pollution sources in Factor 1 of our study. Aneja et al. 2008 [14] reported that ammonia is found in industrial gas emissions or natural sources that evaporate and become particulate matter, and then descend with precipitation and enter surface water. Urbanized areas with high population density in Taiwan also show that domestic wastewater is a major contributor to river water pollution because of the levels of BOD and DO, which show a strong correlation with coliform levels that are associated with domestic wastewater [27]. Runoff from other rivers can be due to flash floods that often occur in Southern Taiwan following typhoons throughout the year. For example, in August 2009, Taiwan experienced the worst floods in 50 years after Typhoon Morakot struck almost the entire southern region. Yang et al. 2012 [32] analyzed the impact of climate change on river water quality in the southern area of Taiwan. High amounts of sediments and debris flowed into the Erren River basin because of the high concentration of suspended sediments in the river, which in turn caused the failure of wastewater treatment plants. Therefore, the river received significantly higher SS, BOD, and ammonia loads from farms and domestic wastewaters. During the dry season, the evapotranspiration rate increase, which may contribute to the increased water salinity. However, during the wet season, precipitation increases and runoff from other tributaries brings SS or nitrate content to the river. Therefore, we assume that climate conditions are one of the factors affecting water quality in rivers.

This study only explored pollution sources that were identified and considered using multivariate statistical analysis of water quality data for all seasons. However, pollution levels vary every season. Therefore, further study is necessary to analyze in detail how different seasons can affect the water quality level. In addition, some water quality variables might be affected by soil types, geological conditions, terrain, and anthropogenic pollution sources [33]. Further work is necessary to determine if these potential sources do significantly impact the rivers in Taiwan.

Our study found that the levels of heavy metal contamination in the Erren River are classified as among the highest in Taiwan. Since the 1970s, the development of a scrap metal industry along the Sanyegong River (a tributary of the Erren River) has severely polluted the river sediment with metals [34]. Chen et al. 2004 [35] determined that concentrations of Fe, As, Cd, Zn, Hg, and Cu in the Erren River were higher than those in other rivers, and that Cu levels exceeded the standard limit. This high heavy metal contamination problem has affected the river ecology and biota. They reported that the highest concentrations of Fe, Zn, Cu, and Mn in muscles were found in tilapia, striped mullet, large-scaled mullet, and milkfish. The highest concentrations of As and Hg were found in striped mullet and Indo-Pacific tarpon. The highest concentrations of Fe, Hg, and Cd were found in the livers of large-scaled mullet, while striped mullet had the highest concentrations of Zn, Cu, and As. Our data in 2002 revealed that As, Cu, Hg, and Zn levels in the Erren River were the highest compared to the other years, indicating that the high levels of the mentioned heavy metals may have affected the biota. However, the trend of heavy metals in the Erren River has been decreasing since the Taiwanese government started a river restoration program in 2002. The restoration program formed an implementation team that united the Water Resources Agency, Industrial Development Bureau, Construction, and Planning Agency, Council of Agriculture, Tainan City Government, Kaohsiung City Government, the river patrol team, and other units to make joint efforts toward improving the water quality of the Erren River. Through the combined efforts of the government and private entities over the long term, the Erren River's water quality is continuing to improve.

5. Conclusions

In this study, cluster analysis was successfully utilized to classify the water quality of 14 Taiwan rivers and PCA–MLR was conducted to determine the possible pollution sources for the most polluted river in Taiwan. According to the cluster analysis, the most severe water quality pollution problem can be found in the Erren River in Southern Taiwan. According to the PCA–MLR results, 62.3% of water pollutants in the Erren River were contributed by ammonia and TP as the first factor; DO, BOD, COD, nitrite, and coliform as the second factor; and conductivity, nitrate, and SS as the third factor. An estimated 72% of the first factor was found to be from industrial emission, 16% from domestic black water, and 12% from natural sources and runoff from another tributary.

Water quality monitoring programs generate complex multidimensional data that require multivariate statistical treatment for analysis and interpretation to obtain better information about the quality of surface water. Such information can help environmental managers make better decisions regarding action plans. The management of domestic and industrial wastes should strive for low accumulation in rivers to minimize environmental degradation. This objective can be achieved by installing proper treatment methods for municipal and industrial wastewater before being released to the environment.

Supplementary Materials: The following are available online at http://www.mdpi.com/2073-4441/10/10/1394/s1, Table S1: Water Quality and Heavy Metal Monitoring Methods, Table S2: Taiwan EPA water quality standards for heavy metal content in surface water, Table S3: River Pollution Index Descriptive Statistic Table, Figure S1: Laojie river map with industrial area and total population data; Figure S2: Heavy metal distribution in Laojie River (sampling locations S1-S7 are in order from upstream to downstream)

Author Contributions: Data curation, C.-H.L.; Supervision, S.-H.O.; Visualization, M.S.A.P.; Writing—original draft, M.S.A.P.; Writing—review & editing, M.S. and Y.-C.W.

Funding: This research was funded by Taiwan Environmental Protection Administration with this following grant number: [EPA-106-U101-02-A121, 2017]; [EPA-106-E3S4-02-07, 2017]; and [EPA-107-E3S4-02-03].

Acknowledgments: We would like to thank the Environmental Protection Administration and the Executive Yuan R.O.C (Taiwan) for providing research data.

Conflicts of Interest: The authors declare no conflict of interest.

References

1. Wang, X.; Zhang, L.; Cai, Y. Heavy metal pollution in reservoirs in the hilly area of southern China: Distribution, source apportionment and health risk assessment. *Sci. Total Environ.* **2018**, *634*, 158–169. [CrossRef] [PubMed]
2. Huang, F.; Wang, X.; Lou, L.; Zhou, Z.; Wu, J. Spatial variation and source apportionment of water pollution in Qiantang River (China) using statistical techniques. *Water Res.* **2010**, *44*, 1562–1572. [CrossRef] [PubMed]
3. Elhatip, H.; Hınıs, M.A.; Gülbahar, N. Evaluation of the water quality at Tahtali dam watershed in Izmir-Turkey by means of statistical methodology. *Stochastic Environ. Res. Risk Assess.* **2007**, *22*, 391–400. [CrossRef]
4. Kolovos, A.; Christakos, G.; Serre, M.L.; Miller, C.T. Computational bayesian maximum entropy solution of a stochastic advection-reaction equation in the light of site-specific information. *Water Resour. Res.* **2002**, *38*, 1–17. [CrossRef]
5. World Bank. *Riding the Wave: An East Asian Miracle for the 21st Century*; The World Bank: Washington, DC, USA, 2017.
6. Tsai, D.H.A. The effects of dynamic industrial transition on sustainable development. *Struct. Chang. Econ. Dyn.* **2018**, *44*, 46–54. [CrossRef]
7. Administration, T.E.P. *Taiwan River Annual Report*; Taiwan Environmental Protection Administration: Taipei, Taiwan, 1998.
8. Chen, Y.-C.; Yeh, H.-C.; Wei, C. Estimation of River Pollution Index in a Tidal Stream Using Kriging Analysis. *Int. J. Environ. Res. Public Health* **2012**, *9*, 3085–3100. [CrossRef] [PubMed]
9. Liou, S.-M.; Lo, S.-L.; Hu, C.-Y. Application of two-stage fuzzy set theory to river quality evaluation in Taiwan. *Water Res.* **2003**, *37*, 1406–1416. [CrossRef]

10. Hsieh, P.-Y.; Shiu, H.-Y.; Chiueh, P.-T. Reconstructing nutrient criteria for source water areas using reference conditions. *Sustain. Environ. Res.* **2016**, *26*, 243–248. [CrossRef]
11. Lai, Y.C.; Tu, Y.T.; Yang, C.P.; Surampalli, R.Y.; Kao, C.M. Development of a water quality modeling system for river pollution index and suspended solid loading evaluation. *J. Hydrol.* **2013**, *478*, 89–101. [CrossRef]
12. Chiu, Y.-W.; Yeh, F.-L.; Shieh, B.-S.; Chen, C.-M.; Lai, H.-T.; Wang, S.-Y.; Huang, D.-J. Development and assays estradiol equivalent concentration from prawn (p-EEQ) in river prawn, Macrobrachium nipponense, in Taiwan. *Ecotoxicol. Environ. Saf.* **2017**, *137*, 12–17. [CrossRef] [PubMed]
13. Elsayed, E.A. Using QUAL2K Model and river pollution index for water quality management in Mahmoudia Canal, Egypt. *J. Nat. Resour. Dev.* **2014**. [CrossRef]
14. Aneja, V.P.; Blunden, J.; James, K.; Schlesinger, W.H.; Knighton, R.; Gilliam, W. Ammonia assessment from agriculture: U.S. status and needs. *J. Environ. Qual.* **2008**, *37*, 515. [CrossRef] [PubMed]
15. Morales, M.M.; Marti, P.; Llopis, A.; Campos, L.; Sagrado, S. An environmental study by factor analysis of surface seawaters in the Gulf of Valencia (Western Mediterranean). *Anal. Chim. Acta* **1999**, *394*, 109–117. [CrossRef]
16. Liou, S.-M.; Lo, S.-L.; Wang, S.-H. A generalized water quality index for Taiwan. *Environ. Monit. Assess.* **2004**, *96*, 35–52. [CrossRef] [PubMed]
17. Li, B.; Yang, G.; Wan, R.; Hormann, G.; Huang, J.; Foher, N.; Zhang, L. Combining multivariate statistical techniques and random forests model to assess and diagnose the trophic status of Poyang Lake in China. *Ecol. Indic.* **2017**, *83*, 74–83. [CrossRef]
18. Shrestha, S.; Kazama, F. Assessment of surface water quality using multivariate statistical techniques: A case study of the Fuji river basin, Japan. *Environ. Model. Softw.* **2007**, *22*, 464–475. [CrossRef]
19. Singh, K.P.; Malik, A.; Sinha, S. Water quality assessment and apportionment of pollution sources of Gomti river (India) using multivariate statistical techniques—A case study. *Anal. Chim. Acta* **2005**, *538*, 355–374. [CrossRef]
20. Singh, K.P.; Malik, A.; Mohan, D.; Sinha, S. Multivariate statistical techniques for the evaluation of spatial and temporal variations in water quality of Gomti River (India)—A case study. *Water Res.* **2004**, *38*, 3980–3992. [CrossRef] [PubMed]
21. Chen, H.; Teng, Y.; Yue, W.; Song, L. Characterization and source apportionment of water pollution in Jinjiang River, China. *Environ. Monit. Assess.* **2013**, *185*, 9639–9650. [CrossRef] [PubMed]
22. Taylor, R. Interpretation of the correlation coefficient: A basic review. *J. Diagn. Med. Sonogr.* **1990**, *6*, 35–39. [CrossRef]
23. Pett, M.A.; Lackey, N.R.; Sullivan, J.J. *Making Sense of Factor Analysis: The Use of Factor Analysis for Instrument Development in Health Care Research*; SAGE Publications: Dartford, UK, 2003.
24. Yang, L.; Mei, K.; Liu, X.; Wu, L.; Zhang, M.; Xu, J.; Wang, F. Spatial distribution and source apportionment of water pollution in different administrative zones of Wen-Rui-Tang (WRT) river watershed, China. *Environ. Sci. Pollut. Res. Int.* **2013**, *20*, 5341–5352. [CrossRef] [PubMed]
25. Kim, G.-H.; Jung, K.-Y.; Yoon, J.-S.; Cheon, S.-U. Temporal and spatial analysis of water quality data observed in lower watershed of nam river dam. *J. Korean Soc. Hazard Mitig.* **2013**, *13*, 429–437. [CrossRef]
26. Carpenter, S.R.; Caraco, N.F.; Correll, D.L.; Howarth, R.W.; Sharpley, A.N.; Smith, V.H. Nonpoint pollution of surface waters with phosphorus and nitrogen. *Ecol. Appl.* **1998**, *8*, 559. [CrossRef]
27. Vega, M.; Pardo, R.; Barrado, E.; Debán, L. Assessment of seasonal and polluting effects on the quality of river water by exploratory data analysis. *Water Res.* **1998**, *32*, 3581–3592. [CrossRef]
28. Islam, M.A.; Islam, S.L.; Hassan, A. *Impact of Climate Change on Water with Reference to the Ganges–Brahmaputra–Meghna River Basin, in Chemistry and Water*; Elsevier: Amsterdam, The Netherlands, 2017.
29. Chinn, D.L. Rural poverty and the structure of farm household income in developing countries: Evidence from Taiwan. *Econ. Dev. Cult. Chang.* **1979**, *27*, 283–301. [CrossRef]
30. Hajigholizadeh, M.; Melesse, A.M. Assortment and spatiotemporal analysis of surface water quality using cluster and discriminant analyses. *Catena* **2017**, *151*, 247–258. [CrossRef]
31. Lee, C.-C.; Jiang, L.-Y.; Kuo, Y.-L.; Hsieh, C.-Y.; Chen, C.S.; Tien, C.-J. The potential role of water quality parameters on occurrence of nonylphenol and bisphenol A and identification of their discharge sources in the river ecosystems. *Chemosphere* **2013**, *91*, 904–911. [CrossRef] [PubMed]
32. Yang, C.P.; Yu, Y.T.; Kao, C.M. Impact of climate change on kaoping river water quality. *Appl. Mech. Mater.* **2012**, *212*, 137–140. [CrossRef]

33. Wu, C.; Wu, J.; Qi, J.; Zhang, L.; Huang, H.; Lou, L.; Chen, Y. Empirical estimation of total phosphorus concentration in the mainstream of the Qiantang River in China using Landsat TM data. *Int. J. Remote Sens.* **2010**, *31*, 2309–2324. [CrossRef]
34. EPA. *Our Erren River—Propaganda Handbook of River Conservation*; Environmental Protection Administration, Executive Yuan: Taipei, Taiwan, 2011.
35. Chen, Y.C.; Hwang, H.J.; Chang, W.B.; Yeh, W.J.; Chen, M.H. Comparison of the metal concentrations in muscle and liver tissues of fishes from the Erren River, southwestern Taiwan, after the restoration in 2000. *J. Food Drug Anal.* **2004**, *12*, 358–366.

MDPI
St. Alban-Anlage 66
4052 Basel
Switzerland
Tel. +41 61 683 77 34
Fax +41 61 302 89 18
www.mdpi.com

Water Editorial Office
E-mail: water@mdpi.com
www.mdpi.com/journal/water

www.ingramcontent.com/pod-product-compliance
Lightning Source LLC
LaVergne TN
LVHW070637170726
843515LV00004B/270

* 9 7 8 3 0 3 6 5 1 8 5 8 9 *